EARTHLIKE PLANETS

Apollo 15 photograph of the Moon, viewed northward over the crater Autolycus and the 50-kilometer-diameter crater Aristillus (near top of image).

EARTHLIKE PLANETS

Surfaces of Mercury, Venus, Earth, Moon, Mars

Bruce Murray
Director, Jet Propulsion Laboratory,
California Institute of Technology

Michael C. Malin
Arizona State University

Ronald Greeley
Arizona State University

W. H. FREEMAN AND COMPANY
San Francisco

Manuscript Editor: Dick Johnson
Designer: Robert Ishi
Production Coordinators: Fran Mitchell and Bill Murdock
Illustration Coordinator: Cheryl Nufer
Artists: John and Judy Waller
Compositor: York Graphic Services
Printer and Binder: The Maple-Vail Book Manufacturing Group

Library of Congress Cataloging in Publication Data

Murray, Bruce C
  Earthlike planets.

  Includes bibliographies and index.
  1. Planets—Surfaces. 2. Moon—Surface.
I. Malin, Michael C., joint author. II. Greeley,
Ronald, joint author. III. Title.
QB601.M86   559.9'2   80-19608
ISBN 0-7167-1148-6
ISBN 0-7167-1149-4 (pbk.)

Printed in the United States of America

9 8 7 6 5 4 3 2 1

CONTENTS

PREFACE

We are all transitory inhabitants of the third planet orbiting the Sun—an average star in an average galaxy containing at least two billion similar stars and probably many billions of other planets. Together we are witness to an unprecedented episode in our species' development, when the capacities to explore Earth's physical surroundings and widely disseminate the results have expanded incredibly fast and effectively. As a result, we are immersed in a planetary information explosion.

Traditional explanations of the nature and history of Earth and the other rocky, Earthlike planets of the inner Solar System—Moon, Mars, Venus, and Mercury—are crumbling under the impact of close-up and direct observations of actual surface phenomena. New insights are developing that link Earth, including the very atoms that compose its sentient beings, with the origin and evolution of those other four planets of the inner Solar System.

A scientific Book of Genesis is emerging, no less dramatic than the literary one; a common planetary environmental history is unfolding that unites the surface histories of Earth and the other inner planets and probably places some conditions on our planetary future.

This book makes available to college undergraduates and to other serious nonspecialists the outlines of the intellectual revolution under way concerning Earth and its kindred planets, especially their surface features, processes, and histories. No equations are used, and reliance is placed on the reader's physical intuition rather than on any previously developed expertise in physical theory. As a consequence, the book should be suitable as a supplementary text in college geology and astronomy courses and also of use in specialized courses covering topics in physical geology, geomorphology, planetary astronomy, volcanology, and planetary science. In order to facilitate diverse potential uses of this book, individual chapters generally contain sufficient background to permit them to be read out of sequence if necessary. General references are included at the end of each chapter for those who wish to explore further.

It is our hope that this book contains a sufficiently accessible overview of the surfaces of the Earthlike planets so that many others who are curious about planetary exploration (for example, readers of *Scientific American, Sky and Telescope, Nature,* or *Science*) will find stimulation and edification from it. A major objective has been to bring the subject of the Earthlike planets into a broad intellectual framework, including past parallels and future possibilities.

Planetary exploration surely is one of the brightest accomplishments of the second half of the twentieth century; if college students and others find that this book serves as a window into some of the contemporaneous additions to the intellectual heritage of humankind, then our purposes will be superbly met.

June 1980   *Bruce Murray*
                                               Michael C. Malin
                                               Ronald Greeley

ACKNOWLEDGMENTS

No single scientist, or even three, can have all the expert knowledge, much less the balanced perspectives, needed to do justice to the present state of knowledge concerning the surfaces of the Earthlike planets. To the extent we have succeeded in producing a useful book, the helpful criticism and suggestions of many of our colleagues have been essential. We wish to acknowledge in particular reviewers of specific portions of the manuscript: Professor Arden Albee, Division of Geological and Planetary Sciences, and Jet Propulsion Laboratory, Caltech; Dr. Michael Carr and Dr. Newell Trask, United States Geological Survey; Dr. Clark R. Chapman and Dr. William K. Hartmann, Planetary Science Institute; Dr. Fraser P. Fanale, Jet Propulsion Laboratory, Caltech; Dr. Donald E. Gault, Murphys Center of Planetology; Professor John E. Guest, University of London, England; Professor William M. Kaula, Department of Earth and Space Sciences, and Institute of Geophysics and Planetary Physics, University of California, Los Angeles; Professor Robert B. Leighton, Division of Physics, Mathematics, and Astronomy, Caltech; Dr. Peter H. Schultz and the late Dr. Thomas R. McGetchin, Lunar and Planetary Institute; Professor Robert P. Sharp, Division of Geological and Planetary Sciences, Caltech; Professor Robert G. Strom, Lunar and Planetary Laboratory, University of Arizona; and Dr. George W. Wetherill, Department of Terrestrial Management, Carnegie Institution of Washington.

In addition, Dr. Carl Sagan of Cornell University, Dr. Thomas Mutch of Brown University, and Mr. Paul D. Spudis of Arizona State University reviewed the entire book-length manuscript and provided important literary and technical criticism.

We wish to acknowledge the joint support of Bruce Murray's initial efforts by the John Simon Guggenheim Memorial foundation and Caltech. All three authors, while not directly supported by the National Aeronautics and Space Administration during the writing of this book, wish to call attention to the rapid development of the subject of comparative planetology stimulated by NASA, in particular, by the foresight of Stephen Dwornik, formerly of NASA Headquarters.

Graham Berry of Caltech was of assistance in ironing out some of the more glaring stylistic clashes amongst the three authors' early drafts. Dick Johnson of W. H. Freeman and Company has provided superb editorial assistance and guidance in the final editing process.

Finally, we wish to express our sincere appreciation to Lorna Griffith of Caltech for her good-humored, capable, persistent, and patient efforts to pull together draft after draft of chapters, and finally a book, from our inconsistent and uneven efforts. If virtue is its own reward, then she is wealthy indeed!

EARTHLIKE PLANETS

1

AN INTELLECTUAL
REVOLUTION

AN INTELLECTUAL REVOLUTION

Humankind has broken the chains of gravity and taken the first tentative flights from its ancestral home, beginning with orbital flights about the Earth and culminating in the first visit to the Moon. The extraordinary accomplishments that mark the age of space exploration have permanently altered our view of ourselves and our potential.

Paralleling the revolution in human perspective inspired by the grandeur and majesty of human space flight has been a revolution in our view of Earth (Fig. 1.1) and its kindred planets within the inner Solar System. The Moon, Mercury, Venus, and Mars have been surveyed by sophisticated robots. Related studies of all kinds—laboratory analyses, Earth-based telescopic observations, and theoretical analyses—have been enormously expanded to accompany close-up and direct observations by space probes.

Orbiting spacecraft have taken spectacular photographs, and entry probes and landers have made esoteric surface and atmospheric measurements on the inner planets—the *Earthlike* planets. The information so acquired is being combined and compared with the knowledge gained from intricate analysis of meteorites and lunar samples. The results truly challenge our intellectual limits as our place in the evolution of the Solar System—indeed, in the entire physical universe—is revealed in new and unimagined ways.

The exploration of our planetary neighbors has provided extraordinary fare for our insatiable curiosity about our own environment; it has provided specific findings of importance to understanding environmental problems on Earth. Surely it will

Figure 1.1
Earth as viewed by the last humans to visit the Moon (Apollo 17). When seen from space, Earth displays three prominent colors: the clouds of the atmosphere are white, the continents are brown or reddish-tan, and the oceans are blue-black.

provide the basis for future utilitarian activity, just as the expansion of terrestrial geology at the end of the nineteenth century from a continental scale to a global scale provided the basis for new views of the Earth. So, likewise, do planetary geology and exploration now represent an intellectual and societal challenge of what is truly the next frontier.

The logic of geology—terrestrial or planetary—is tripartite. First, the present state of the body, and especially its surface, must be understood. Second, the processes that operate now and have operated in the past must be comprehended. Third, one can attempt to reconstruct the history of the surface of a planet by reconciling the present state of the planet with the variety, scale, and sequence of the processes that have operated on it.

Thus our task is to examine a selection from the enormous amount of data gathered during the space age about the surfaces of our planetary neighbors. In our examination, we shall compare what is known about the Earth with what has been learned about the rest of the inner Solar System. The very earliest part of the record of planet formation is, of course, the least preserved, since it has been obscured by so many later events. Nevertheless, the results obtained by the many studies that space exploration has stimulated over the past two decades have caused scientists to revise considerably their views of planetary origin. It therefore seems appropriate to begin our study by showing just how the scientific view of the origin of the Earth and other inner planets has changed.

IN THE BEGINNING—PAST VIEWS

Since the seventeenth century, scientists have imagined that the planets somehow formed along with the Sun from a disk of gas and dust. The evolution of this embryonic Solar System (called the *solar nebula*) remains difficult to characterize at the point when the planets actually began to form. Even so, our understanding of how Earth—and life—ultimately evolved from the solar nebula has grown considerably.

Planetary Origin—A Rare or Common Stellar Phenomenon?

Until recently, it was so difficult to imagine all the complicated steps in the formation of large solid planets from dust and gas that the very process itself was considered by many to be unusual, perhaps even unique to our Solar System. Even the nearest stars are so distant that we cannot detect directly any planets that may be orbiting them. If planets are deemed to be rare in the Universe, then clearly the concept of life—especially intelligent life—elsewhere in the Universe must be more the province of science fiction writers than scientists.

One of the primary results of space age research has been to demystify the process of planet formation, and thus the origin of life. Hence the growth in our understanding of the origin and evolution of the Earth and its neighbors adds greatly to the likelihood that other stars may be orbited by planets with surfaces like that of the Earth, including the presence of liquid water. The increased expectation that there are other habitable planets raises the hopes of many scientists that we are not alone in this Galaxy and stimulates consideration of how we might find direct evidence of the existence of extraterrestrial intelligence.

Cold, Homogeneous Accretion

There were many pre-space age speculations about a mysterious epoch during which smaller objects began to accumulate and stick to larger objects, gradually forming planet-sized bodies (Fig. 1.2). These speculations led to the prevailing notion that Earth—and presumably the other objects in the Solar System—was formed by accretion of numerous small bodies whose chemical composition was, on the average, like that of the entire Solar System, although modified somewhat by distance from the Sun. This hypothetical process is termed *homogeneous accretion*. In addition, the formation process was generally (but not universally) thought to be one in which the Earth remained fairly cold during accretion and did not heat up sufficiently until after the entire mass had accumulated to become molten and undergo chemical differentiation into core, mantle, and crust.

It was generally thought also that the Earth initially must have had a massive, primary atmosphere composed, like that of the Sun, mostly of hydrogen and helium. Yet the air we breathe

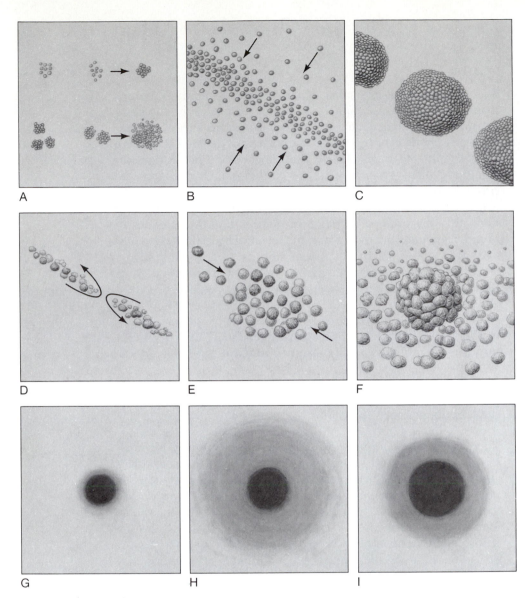

Figure 1.2
Planets are believed to have started forming when dust grains of the solar nebula collided and stuck to one another, forming ever larger clumps **(A)**. The clumps fall toward the midplane of the nebula **(B)** and form a diffuse disk there, aided by the drag of the gas still present in the nebula. Gravitational instabilities collect this material into millions of bodies of asteroid size **(C)**. These in turn form clusters **(D)**. When clusters collide and intermingle **(E)**, they presumably coagulate into solid cores, perhaps with some bodies going into orbit around the cores **(F)**. Continued accretion and consolidation may create a body the size of one of the inner planets **(G)**. If the core gets larger, it may concentrate gas from the nebula gravitationally **(H)**. A large enough core may make the gas collapse into a dense shell that constitutes most of the planet's mass, as in the outer planets at present **(I)**. [After "The Origin and Evolution of the Solar System" by A.G.W. Cameron. Copyright © 1975 by Scientific American, Inc. All rights reserved.]

on Earth is vastly deficient in hydrogen and helium. Scientists therefore hypothesized that Earth's primary atmosphere was completely removed before the present atmosphere and the oceans began to form. Consequently, the Earth's present atmosphere was referred to as "secondary." Some young stars exhibit gigantic flares at periods in their history when vast amounts of hot gases are violently ejected. Some such phase was speculated to be responsible for literally "blowing away" the primary hydrogen and helium atmospheres of the inner planets at a very early time.

Heat-generating radioactive elements would have been dispersed uniformly throughout the primitive Earth according to this early view of homogeneous accretion. The rapid decay of short-lived radioactive elements, like one isotope of aluminum (atomic mass 26), was believed to have promptly heated the Earth sufficiently to cause iron (bound up in the objects that collided with the growing Earth) to melt, separate, and settle toward the center, thereby forming what is now the Earth's core. Most of the rest of the planet was deemed to be composed of the residual silicate slag, which we now recognize as the mantle of the Earth. In addition, a thin, low-density silicate crust was formed that is not only rich in aluminum and silicon but also contains the highly radioactive long-lived isotopes of the elements uranium, thorium, and potassium, all of which have a high chemical affinity for low-density aluminum silicate minerals. Continued chemical differentiation of a nearly molten Earth (as a consequence of the same internal radioactive heating augmented by additional heat released by the formation of the core) was presumed to have caused volatile compounds, such as water and carbon dioxide, to be liberated from the mineral matrices in which they were chemically and physically bound at the time of accretion. Substantial portions of those volatiles were presumed to have made their way gradually to the surface through volcanic emanations to form the present atmosphere and hydrosphere.

Early Oceans and Life

Even before the space age, there was lively disagreement over how rapidly the Earth's oceans and atmosphere formed. Some geologic evidence suggested that most, if not all, of the Earth's atmosphere and oceans formed very early; yet numerous scientists argued to the contrary that transfer to the surface of the necessary water, carbon dioxide, and other volatiles extended over a long period.

How is the origin of life connected to the origin of the oceans? It was hypothesized that a "soup" of organic molecules, based on hydrogen, ammonia, and methane gases, accumulated in the early oceans and that accidental combinations of the complex life-building molecules—nucleotides and amino acids—led by chance to even more complex ones capable of self-replication. Such molecules evolved to form the first microorganisms, nourished by the rich, raw, pre-biotic organic materials available to them. Microorganisms even more primitive than blue-green algae were believed to be among the first to have formed by this process, and presumably constituted the basis of most later life forms. It was also believed that free oxygen was probably only a minor constituent of the original atmosphere and that subsequent plant life eventually produced most of the oxygen now in our atmosphere.

Thus the Earth was regarded as having formed in a fairly uniform manner from the solar nebula, the initial stage of condensation and accretion being followed by chemical differentiation along with the gradual release of volatiles. We and all other living earthlings were believed to be the product of uniform chemical and then biological evolution that accompanied the development of the hydrosphere.

IN THE BEGINNING—CURRENT VIEWS

How do our views now compare with those of the pre-space era? Particularly, how do they differ with respect to the early processes that formed the planets? We now have a richer range of ideas, and many of the earlier ones have been discarded. But we still enter the realm of speculation when trying to probe back through time to the origins of the planets.

Hot, Heterogeneous Accretion

First of all, scientists now recognize that chemical differentiation characterizes not only Earth, but the Moon, Mercury, Mars, and probably Venus. In addition, two new ideas have gained increasing acceptance since the space age began. First, the planets accumulated while hot, not cold. Second, the Earth and the other planets perhaps did not originate as homogeneous, solid planets and then become differentiated later; rather, their differentiation may have taken place *during* the accretion process. The present radial segregation into distinct chemical zones may be the consequence of *heterogeneous* accretion. In that view, it is imagined that in the early phases, when the gas in the solar nebula was very hot, perhaps 1000°C, mainly dense iron-rich compounds precipitated and accumulated in proto-planets to form what would become the iron-rich planetary cores. As the temperature of the nebula dropped, lower-density silicate minerals began to condense and be swept up by the proto-planets. (Thus, for example, the composition of the Earth's early mantle might have been similar to that of today's mantle.)

Most important, space age evidence indicates that at the time they formed, the Earth and the other planetary objects were extremely hot—even molten throughout. Hence chemical differentiation must have been taking place while the planets were being formed. Cold, homogeneous accretion is no longer a viable theory; the story of planetary genesis has been rewritten. The change from a cold birth to a hot one has profound implications for planetary evolution. Rather than a gradual increase of internal temperature throughout geological time, the planets may actually have been cooling since the time of accretion, with the smallest planets, the Moon and Mercury, having become inactive at the surface billions of years ago. Earth and Venus, by this view, are still "barely congealed," and their interiors continue to interact with their surfaces. Mars is viewed as an intermediate case.

Curiously enough, there are a few tiny pieces of silicate minerals that do seem to have survived from the solar nebula without melting. Tiny inclusions in some meteorites preserve fossil chemical and isotopic "fingerprints" that date from before the origin of the Solar System—before there was light from the Sun! Thus, even while the present intellectual revolution restructures and unifies our understanding of the birth of the planets, new clues to future revolutionary ideas may be emerging that will help us to trace our origins back even further in time.

Earth's Volatiles—Not from a Secondary Atmosphere

If chemical differentiation occurred during—or even before—accretion, when did the volatiles evolve? According to the concept of heterogeneous accretion, they could have been contained only in the impacting bodies that arrived during the later stages of accretion. The temperature in the solar nebula had to have decreased enough for volatile elements to become incorporated in solids. According to this view, volatiles never could have been scattered homogeneously throughout the entire Earth, but rather were confined to the outer regions, which were the last to form. Even if the impacting objects were homogeneous chemically throughout accretion, their contained volatiles must have been released almost immediately upon incorporation into the proto-Earth. Indeed, one new view is that the growing Earth came to be surrounded by a cloud of steam and other volatiles that finally condensed to form the oceans and atmosphere once accretion was complete and the surface cooled sufficiently.

But even more extraordinary speculations have been stimulated by space age results. It has been conjectured that volatile-rich impacting objects that contributed to the outermost layers of the Earth and its neighboring Earthlike planets were distinct in origin from those that accumulated to form the main mass of the planets. These hypothetical volatile-rich bodies conceivably arrived late (perhaps half a billion years after the beginning of accretion) from orbits that reached farther out from the Sun than Earth's—perhaps even out beyond the orbit of Jupiter. Such "trans-jovian" objects have long been inferred to be the source of comets, which are known to contain large amounts of organic material and water.

Certain kinds of meteorites also contain organic material and volatiles, and may themselves be old cometary debris swept up by Earth. The chemical and isotopic similarity of the rare gases in the atmospheres of Earth, Mars, and Venus to those in the meteorites rather than to average solar abundances suggests a distinct and common history for the volatiles on the Earthlike planets and in the meteorites. The volatiles from both sources probably were modified together in some way very early in the history of the solar nebula. Moreover, meteorites originate farther from the Sun than Earth's orbit, suggesting that perhaps the Earth's volatiles also originally condensed out beyond Mars, or Jupiter, or even farther. The great bulk of the Earth's solid material surely accumulated from objects condensed at the same distance from the Sun as the Earth—that is, in similar accretionary orbits. But it is at least conceivable, though still not the most likely circumstance, that the Earth was bombarded late in the history of its formation by distinctly different objects that had formed farther out in the Solar System and were much richer in volatiles. In any case, the Earth and the Earthlike planets probably never had a "primary" atmosphere of hydrogen and helium. The current atmospheres of Earth, Mars, and Venus probably are derived only from the near-surface portions of those bodies, and the oceans and atmosphere of Earth very likely came into full existence at least 4 billion years ago, rather than being gradually built up over geologic time.

How Did Life Form on Earth?

Even the origin of life seems less easily explained as the result of our new knowledge. For example, it has been shown by laboratory experiments that the critical amino acid building blocks of life could have formed by a freeze/thaw process at an early stage in Earth's history. This hypothetical mechanism provides one possible alternative to the traditional view that life evolved from an organic "soup" that accumulated in the early oceans. This is of interest because it is likely on astrophysical grounds that the luminosity of the Sun has been increasing slowly over geologic time, tending to heat up the Earth; Earth's surface may have been characterized at an early stage by frozen accumulations of volatiles. Alternatively, reduced gases, such as ammonia, hydrogen, and methane, probably present before the accumulation of oxygen in the early atmosphere, may have absorbed enough radiant heat from the surface to make the Earth even warmer then than it is at present, a condition that would be consistent with the original organic "soup" hypothesis of the origin of life.

Yet, what if the volatiles necessary to form the atmosphere, hydrosphere, and life arrived all at once from a more distant part of the Solar System after accretion of the solid planet? Then it would no longer be *required* for life to have formed on the Earth itself. Nonbiologically synthesized amino acids survive meteoritic transport to Earth today. Could the simplest life forms likewise have survived during the torrential flux of impacting objects at, or shortly after, the end of Earth's accretion? (This speculation merely transfers the key initial step in the evolution of life to some other environment within the Solar System.) Thus not all the critical steps from pre-biotic organic compounds to the simplest life forms necessarily had to take place on Earth, although that definitely still is the most plausible circumstance.

THE FAMILY OF EARTH—CHANGING GENEALOGY

Moon—Uncertain Parentage

In many ways the Moon was enigmatic to both the astronomer and the geologist before space exploration began. Hypothetically, it could have formed separately from the Earth and then been captured early by the Earth's gravitational field. Or the Earth and the Moon could have formed as double planets from the same vortex in the solar nebula. Or possibly the Moon was produced by fissioning of the Earth after its formation. Indeed, when the space age began, some textbooks still referred to a nineteenth-century hypothesis that the formation of the Moon produced the Pacific Ocean basin.

Speculations as to what the Moon was like were legion. Perhaps the most widely accepted were advocated by Nobel Prize winner Harold Urey, who argued that the Moon accreted while cold, had never differentiated chemically, and might well still retain many attributes of that original chemical formation. The heavily cratered lunar topography (Fig. 1.3) was almost universally believed to be left over from the end of the accretionary phase. Thus it was entirely plausible that when samples from the Moon eventually were returned to the Earth for laboratory analysis, scientists would be sampling the raw, unmodified material from which the Moon—and perhaps even Earth—had formed.

Many thought that conspicuous, smooth, dark fillings of the mare basins were of volcanic origin. Alternative arguments

Figure 1.3

Observatory photograph of the Moon showing dark *maria* and light-toned highlands, or *terrae* regions. The heavy cratering of the terra regions indicates their old age; they date back to the last stage in the formation of the Solar System, when the inner planets, including Earth's Moon, were bombarded by asteroid-size objects. The maria are composed of younger lava flows. This image is a composite of two photographs (note the shadow zone, center, where the two halves were joined.) [Lick Observatory photograph.]

abounded—even that the maria were accumulations of cosmic dust! In fact, there was no decisive evidence that *any* lunar feature was actually the result of volcanic processes. Some scientists felt that everything could be explained by impact processes of one kind or another.

Finally, many recognized that the Moon had been generally quiet throughout most of the 4.5 billion years of geologic time, certainly for a billion years or more. But the age of lunar surface features could not be determined without speculative assumptions concerning the frequency with which lunar impact craters had formed and the relationship of those impacts to the rate at which small impact craters have formed on Earth recently.

Surprisingly, despite the enormously extensive investigation of the Moon from space and the opportunity to examine many samples of it in sophisticated terrestrial laboratories, the Moon's origin is still significantly uncertain. Strong chemical and isotopic similarities have been discovered between many of the elements that constitute both the Moon and the Earth. It now seems much less likely that the Earth captured the Moon from some very different location in the Solar System where accumulation conditions must have been quite different. Compared with terrestrial rocks, however, Moon rocks definitely indicate a higher-temperature segregation phase very early in their formation. The matter that formed the Moon apparently was initially heated to a higher temperature than the average material that formed the Earth. This fact, along with the invariant laws of gravity that controlled the motions of the Earth and the Moon, pose severe constraints on mechanisms for lunar origin. No current theory explains entirely satisfactorily all the facts that we now know about the Moon.

However, the Moon clearly did *not* accumulate while cold, it *is* a highly differentiated object, the dark mare *are* volcanic. Furthermore, it has a crust thicker than the Earth's. This crust probably survives from the earliest period of the Moon's solidification as a planet nearly 4.5 billion years ago and manifests a definite asymmetry in thickness; the crust of the farside (the hemisphere permanently hidden from Earth) is thicker than that of the nearside.

The study of the Moon during the Apollo program produced a great surprise concerning the age of the scarred topography previously considered to be left over from the accretionary process. That cratered surface was actually produced long after the time of accretion. The largest basins were definitely formed hundreds of millions of years after the planet accumulated. Furthermore, the ages of formation inferred from intricate study of the isotope ratios of mineral grains in returned samples, as well as from photogeologic studies, indicate a relatively rapid cessation of bombardment. It has not been possible to resolve confidently whether that bombardment episode was the "tail end" of accretion or was an entirely distinct episode, corresponding, for example, to an influx of objects originating farther out in the Solar System. Either way, this terminal period of heavy bombardment seems to have occurred not just on the Moon but on Mercury and Mars as well, and, by analogy, surely must have played an important role on Earth. Thus the oldest surface event planetary geologists have confidently extracted so far from the correlation of the geological records of the inner planets has not been the end of planetary formation nearly 4.5 billion years ago, but instead this distinct terminal phase of massive bombardment that affected the entire inner Solar System about 4 billion years ago.

Mars—A Case of Mistaken Identity

Before the space age, Mars was widely believed to be the planetary twin of the Earth. Both planets exhibit the same obliquity and therefore have nearly identical seasonal patterns. Mars' daily rotation rate is virtually the same as the Earth's. The brilliant white caps that migrate seasonally from each of the martian poles (Fig. 1.4) were universally believed to be composed of water ice. Now we know from better theory that the similarities in spin rate and obliquity are completely fortuitous; and from Mariner 7 observations in 1969 that the white frost seen from Earth is not water—it is frozen carbon dioxide! But in the early 1960s, early martian oceans were considered plausible, if not probable, presumably having evaporated during geologic history. Indeed, the investigators of the first space probes to Mars in 1965 were prepared to look for folded sedimentary mountain belts and other features that might be associated with ancient oceans.

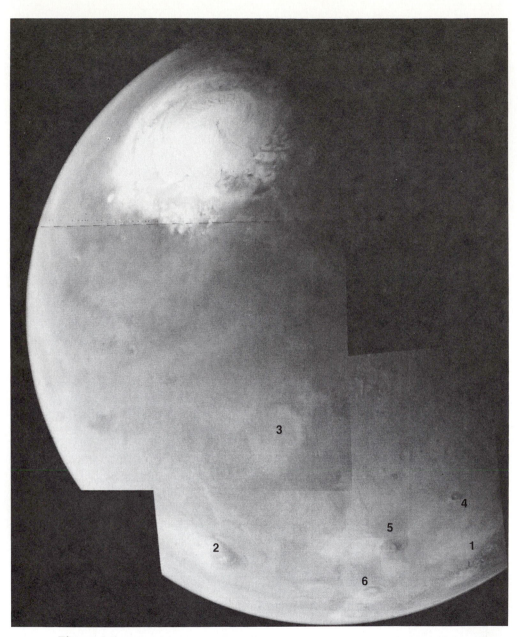

Figure 1.4
Mariner 9 view of Mars. This three-frame mosaic shows the north polar region, identified by the seasonal white cap of carbon dioxide frost (upper left), several of the large volcanoes discovered during the Mariner 9 mission in 1971, and part of *Valles Marineris,* the "Grand Canyon" of Mars (1). The volcanoes that can be seen are *Olympus Mons* (2), *Alba Patera* (3), *Tharsis Tholus* (4), *Ascraeus Mons* (5), and *Pavonis Mons* (6).

Because so many attributes of Mars and the Earth were considered to be quite similar, life, too, was considered plausible on Mars. Indeed, many who studied that planet believed that its seasonal variations of light and dark markings were likely to be plant life going through an Earthlike seasonal cycle. Mariner 9 in 1971 and 1972 demonstrated convincingly that these variations in markings are not manifestations of life, but instead are due to dust storms and their effects on the surface and atmosphere.

The widespread belief that Mars had an atmosphere similar to the Earth's was exemplified by the initial phase of the Soviet space program. The unsuccessful spacecraft Zond II (1962) was very likely a Soviet Mars entry-probe attempt, although the Soviets have never acknowledged the purpose of that mission. It evidently was intended to deliver a probe to the surface by means of a simple parachute system. We now know that because atmospheric pressure on Mars is so much lower than that on the Earth (one-half of one percent), this type of entry system would have failed catastrophically. Indeed, the U.S. would likely have made the same mistake were it not for delays in starting its Mars program. The true nature of the thin atmosphere was confirmed by the fading of radio signals from the U.S. flyby Mariner 4 in 1965 as it passed behind the planet.

Nothing we have learned about Mars appears inconsistent with the hypothesis that it formed by heterogeneous accumulation and may once have been entirely molten. But Mars, we now recognize, surely has not had the same atmospheric history as the Earth. The amount of volatiles that have been present on the surface seems to be no more than a few percent that of the Earth per unit surface area. Furthermore, many of the erosional features suggest that some volatile material was originally contained in the rocks of the outer part of the crust and later liberated to the atmosphere through erosion and crustal fracturing. Finally, the survival there of large Moonlike craters, albeit subdued in character, is evidence that Mars could not have possessed either oceans or an Earthlike atmosphere for at least the past 3 or 4 billion years.

Thus the original concept that Mars was the Earth's "twin" with a very similar history—including the formation of oceans and perhaps an analogous origin of lifeforms—now seems nullified. Mars clearly has a distinctive, unearthly history that must be understood in its own terms rather than by simple analogy to the Earth.

Venus—An Unusual Sibling

A similar lack of knowledge of even the bounds of the possible types of atmosphere and surface conditions on Venus was illustrated by the Russian probe Venera 4 in 1968, which was the first space vehicle to reach the atmosphere of that planet. Verena 4 entered the atmosphere and provided crucial first measurements, but it failed after recording an atmospheric pressure corresponding to about twenty times that on the Earth's surface. The Soviet designers were so sure this measurement indicated the actual surface pressure that the result was published and maintained for several years. Eventually, however, the Soviets acknowledged that the probe had actually failed at least 26 kilometers above the surface. The true atmosphere on Venus (Fig. 1.5) is a hundred times more dense at the surface than Earth's and hot enough to melt lead.

The rotation of Venus was unknown before the space age, but its direction of rotation was generally presumed to be prograde, like that of the Earth, and its rate of spin perhaps similar to Earth's. Radar results in the early 1960s, however, demonstrated that it spins in the opposite direction from Earth (retrograde) and much more slowly than Earth. The first successful planetary spacecraft of any kind, the U.S. probe Mariner 2, flew by Venus in 1962, searching unsuccessfully for, among other things, the presence of an Earthlike magnetic field. This was an entirely reasonable expectation, because Venus and Earth are of nearly the same size and mass, and hence might be similar in internal constitution. But no such magnetic field was found, and another expectation was shattered. The featureless cloud cover (in visible light) of Venus was presumed to be made up of water droplets. No one even guessed sulfuric acid droplets, which is the generally accepted current view.

Surprisingly, Venus has been proven to be quite similar to Earth in the total amount of carbon dioxide liberated at its surface. Most of the Earth's carbon dioxide now resides in limestones and dolomites, rocks that are common in the upper portion of the crust. Perhaps Venus has gone through a stage in which the heat-retention effect of its atmosphere caused the planet's temperature to rise relentlessly until its oceans evaporated and the steam dissociated.

But perhaps Venus' history never was quite the same as the Earth's. Perhaps the original bulk concentration of volatiles

Figure 1.5
Global view of Venus obtained by Mariner 10 in 1974 as part of that flyby mission to Venus and Mercury. The planet is enveloped by dense clouds that completely obscure the surface. This image was taken through an ultraviolet filter to enhance subtle differences in cloud structure. These and other images show that atmospheric flow is roughly symmetrical between north and south hemispheres.

THE FAMILY OF EARTH—CHANGING GENEALOGY   **19**

never was comparable with that of the Earth because Venus formed closer to the Sun. Either Venus has lost an amount of water equal to a global layer 3 to 4 kilometers deep, or Earth and Venus never were really alike. Future *in situ* isotopic measurements are needed to provide a definitive answer to what really happened on Venus. We shall strive to learn more about the origins of Earth's water—and ourselves—by measurements carried out by automated spacecraft on Venus.

Mercury—A Mini-Earth in Moon's Clothing?

Before the space age, Mercury was so little known that it was hardly of much interest to science. What little data did exist were believed to indicate that its axial (daily) rotation and its (annual) rotation about the Sun were synchronous, just as the Moon's axial rotation is synchronous with its revolution about Earth, so that the same hemisphere always faces the Earth. However, radar observations from Earth in the mid-1960s revealed that there is no such synchrony; Mercury instead rotates three times about its spin axis for every two revolutions about the Sun. In addition, a tenuous atmosphere was thought likely; indeed, it was conceivable to many that Mercury once could have been Earthlike, but the continued action of the strong solar radiation would have destroyed any Earthlike atmosphere. Such a small object rotating so slowly was believed not to have a magnetic field.

Mercury's formation was not discussed very much, but one leading planetary astronomer even suggested that a violent phase of early solar history (which was hypothesized to have removed the primary atmosphere of the planets) may also have been responsible for stripping off most of Mercury's rocky outer mantle, leaving only a thin rocky layer covering an iron core.

The real Mercury, stunningly revealed by the U.S. flyby Mariner 10, exhibits a surface astonishingly similar to that of the Moon (Fig. 1.6), with no atmosphere and no evidence of a bizarre early period when solar emanations could have modified its bulk composition. Yet a miniature but seemingly Earthlike magnetic field was also discovered—indicative more of an Earthlike, rather than Moonlike, interior. Clearly this is an interesting planet! With its surface record of external bombard-

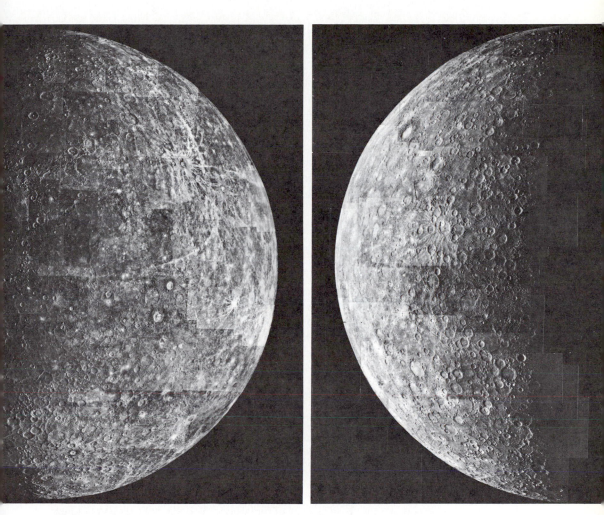

Figure 1.6
The surface of Mercury was seen in detail for the first time through images acquired by
Mariner 10 in 1974. These two mosaics show Mercury during the incoming part (right
side) of the flyby and the outgoing part (left side). Mariner 10 imaged about 54 percent of
the surface of Mercury. The remaining 46 percent remains unexplored.

ment and internal activity, it provides a most important scientific outpost close to the Sun.

In the next chapter, we briefly summarize the global states of the Earthlike planets—their sizes, compositions, and comparative surface anatomy, as would an alien space traveler studying the Earth and its neighbors from orbit. Such a traveler would encounter and measure both similarities and differences among these individual objects, but could scarcely avoid recognizing them as related bodies.

SUGGESTED READING

Bullard, E. "The Origin of the Oceans." *Sci. Am.* **221**(3):66–75 (Sept. 1969). (Offprint No. 880.)

Cameron, A. G. W. "The Origin and Evolution of the Solar System." *Sci. Am.* **233**(3):32–41 (Sept. 1975).

Dickerson, R. E. "Chemical Evolution and the Origin of Life." *Sci. Am.* **239**(3):70–86 (Sept. 1978). (Offprint No. 1401.)

Hibbs, A. R. "The Surface of the Moon." *Sci. Am.* **216**(3):60–72 (March 1967).

Lewis, J. S. "The Chemistry of the Solar System." *Sci. Am.* **230**(3):50–65 (March 1974).

Sagan, C. "The Solar System." *Sci. Am.* **233**(3):22–31 (Sept. 1975).

Schramm, D. N., and R. N. Clayton. "Did a Supernova Trigger the Formation of the Solar System?" *Sci. Am.* **239**(4):124–139 (Oct. 1978). (Offprint No. 3022.)

Shoemaker, E. M. "Geology of the Moon." *Sci. Am.* **211**(6):39–47 (Dec. 1964).

Urey, H. C. "The Origin of the Earth." *Sci. Am.* **187**(4):53–60 (Oct. 1952). (Offprint No. 833.)

Wald, G. "The Origin of Life." *Sci. Am.* **191**(2):45–53 (Aug. 1954). (Offprint No. 47.)

2

THE GLOBAL VIEW

THE GLOBAL VIEW

The landscapes of the Earthlike planets exhibit many similarities. All bear the scars of great collisions with asteroids and comets, and most show evidence of volcanic activity as well. Yet photographs taken from space have revealed that each planet has its own distinctive kinds of terrain and asymmetries. Earth is uniquely the watery planet.

THE MOON

The Face of the Moon

Aside from Earth, the planet we know most about is our own natural satellite, the Moon. Early telescopic observations provided many clues to the geology of the Moon, beginning when Galileo first turned his simple telescope toward the Moon in the early 1600s. Two distinctive types of terrain were immediately apparent (Figs. 2.1, 2.2): the high-albedo (highly reflective), heavily cratered, rugged uplands, named *terrae*; and the low-albedo (dark), smooth, lightly cratered lowlands, named *maria* because of their fancied resemblance to seas. The maria are now known to be vast lava plains that were erupted on the lunar surface from about 3.8 to less than 2.8 billion years ago. These lava plains are composed primarily of the dark, vesicular rock *basalt*, a volcanic rock that is common on Earth and composed of iron- and magnesium-rich minerals. Basaltic lavas are highly fluid, and hence are capable of spreading quickly over large areas.

Figure 2.1
Earth-based view of the full Moon showing dark lava maria, light-toned highland terrae, and numerous craters. The full-Sun illumination in this view enhances differences in reflective properties of the surface, such as the bright rays of ejecta radiating from the fresh craters. Because of the nature of its orbit and rate of spin, the Moon always keeps the same face turned toward Earth, the "nearside." It was not until spacecraft flew to the Moon that the "farside" was seen. At the bottom of the photograph is the bright-rayed crater *Tycho*, near the Moon's south pole. The edges of the Moon in a nearside view are termed "limbs." [Lick Observatory photograph.]

Figure 2.2
View of the eastern limb (seen at upper left here) and part of the farside of the Moon
photographed by the Apollo 16 astronauts. Mare Crisium is the circular dark area in the
upper left, which can be matched with the nearside view in Figure 2.1 for orientation.
Note that most of the farside lacks dark mare regions and is characterized by heavily
cratered terrae. Craters become more apparent toward the terminator (the line separating
the sunlit from the dark portion of a planet) because their relief is emphasized by
shadows.

Although it is now accepted that most lunar craters are the result of impact by meteoroids and comets, this was not fully recognized before the space age, even though more than a century ago there were proponents of an impact origin. The American geologist G. K. Gilbert proposed that even the enormous circular basins, like the one that contains Mare Imbrium, resulted from the impact of large objects—a conjecture now borne out by abundant evidence. Imbrium and the other circular basins constitute the basic architecture of the lunar crust (Figs. 1.3, 5.7, 5.8). Some of these basins are, in addition, filled with the dark lava of the maria. All of the mountain chains on the Moon are concentric blocks of crust that form the rims of these basins.

Synchronized Rotation

The Moon always turns the same face (called the lunar "nearside") toward the Earth (Table 2.1). The synchrony between its axial (daily) rotation and its (monthly) rotation about the Earth is the result of gravitational interactions of the equatorial bulges and other departures from sphericity of both objects. These same mutual gravitational interactions are responsible for most of the oceans' tides. Not until the successful Soviet Luna 3 mission to the lunar farside and the subsequent U. S. Lunar Orbiter missions did it become apparent that the global physiography of the Moon is distinctly asymmetric (Fig. 2.2). Although the large circular basins are randomly distributed over the face of the Moon, very few of the farside basins contain dark mare material. Of the total 17 percent of the lunar surface that is covered by mare lavas, by far the greatest proportion is on the nearside.

Extremes of Temperature

Because the Moon rotates once on its axis for every turn it makes around the Earth, its day is about as long as our calendar month. (Indeed, the "lunar month"—about 28 Earth days in length—was used by primitive societies as a principal means of reckoning time.) Furthermore, the Moon's axis of rotation is

Table 2.1
GLOBAL PROPERTIES OF THE INNER PLANETS

	Mercury	Venus	Earth	Mars	Moon
Mean distance from Sun (millions of kilometers)	57.9	108.2	149.6	227.9	0.3844*
Mean distance from Sun (astronomical units)	0.387	0.723	1	1.524	0.00257*
Period of revolution (Earth days)	88	224.7	365.26	687	27.32*
Rotation period (Earth days)	59	243 retrograde	0.9973	1.026	27.32
Inclination of axis	<2°	3°	23°27′	23°59′	6°41′ relative to orbit about Earth
Eccentricity of orbit	0.206	0.007	0.017	0.093	0.055
Equatorial diameter (kilometers)	4880	12,104	12,756	6787	3476
Mass (Earth = 1)	0.055	0.815	1	0.108	0.01226
Volume (Earth = 1)	0.06	0.88	1	0.15	0.0203
Density (Water = 1)	5.4	5.2	5.5	3.9	3.34
Oblateness	0	0.00001	0.003	0.009	0.0005
Atmosphere (main components)	None	Carbon dioxide	Nitrogen, oxygen	Carbon dioxide	None
Atmosphere (minor components)	None	Noble gases; hydrochloric, hydrofluoric, and sulfuric acids	Noble gases, carbon dioxide	Noble gases, nitrogen	None
Mean temp. at surface (degrees Centigrade)	350° day; −170° night	480°	22°	−23°	107° day; −153° night
Atmospheric pressure at surface (millibars)	<10⁻⁹	90,000	1000	6	0
Surface gravity (Earth = 1)	0.37	0.88	1	0.38	0.16
Mean apparent diameter of Sun as seen from planet	1°22′40″	44′15″	31′59″	21′	31′59″

*Relative to Earth, not Sun.

almost perpendicular to the plane defined by the motion of the Earth about the Sun—that is, the plane of the *ecliptic*. Hence there are virtually no seasons on the Moon. One consequence of this nearly uniform "daily" pattern of sunlight on the Moon's surface is that there are substantial areas near the poles—within craters and other depressions—that are permanently in shadow.

How cold are these perpetually shaded polar surfaces on the airless Moon? At this time, no one is sure, because Earthbound observers cannot view these surfaces. If the Moon's outward heat flow were as high as the Earth's, the temperature at the surface would be −238°C, only 35° above absolute zero. Apollo heat-flow measurements suggest, however, that the lunar surface temperature, in the absence of solar heating, would be even colder! Such localities on the Moon (and Mercury) are among the coldest surface environments within the entire inner Solar System. Because they might retain frozen volatiles from early in the history of the Solar System, these shaded lunar regions offer exciting targets for future exploration.

On the illuminated surface of the Moon, where unfiltered sunlight falls directly on the fine, dark powder that mantles nearly everything, about 93 percent of the Sun's incident light energy is absorbed and, in turn, reradiated thermally as infrared and radio waves. Telescopic measurements of this thermal radiation indicate that surface temperatures reach 130°C—well above the boiling point of water at sea level on Earth—wherever the Sun's rays fall perpendicularly on the surface.

MERCURY

The Harshest Environment

If exposure to noontime Sun at an equatorial locality on the Moon seems harsh by terrestrial standards, consider the circumstances at the surface of Mercury, where the solar insolation is five to ten times more intense. Like the Moon, Mercury has no atmosphere, and is covered with a layer of fine, dark powder. Daytime temperatures reach a maximum of 430°C. At night, radiative cooling of the extremely insulating powdery surface drops temperatures to about − 180°C, as on the Moon.

As is true of the Moon's polar regions, there are permanently shaded areas near the mercurian poles, because the planet's spin axis is similarly perpendicular to the plane of its orbital motion about the Sun. Thus Mercury boasts not only one of the hottest surface environments in the Solar System, but by far the greatest extremes of temperature of any object in the Solar System.

Spin-Orbit Coupling

Mercury's "day" is unique in the Solar System because of the relation between the planet's spin and its orbital motion, called *spin-orbit coupling*. Until 1965, the period of the planet's spin was believed to be exactly equal to its orbital period about the Sun (i.e., its year: 88 Earth days). Thus it was imagined that one hemisphere always faced the Sun and that the other was in perpetual shadow. However, distortions in signals reflected off the rotating planet by very powerful scientific radars on Earth showed that the period of Mercury's spin is significantly less than its 88-day orbital period. In fact, we now know that Mercury's period of rotation on its own axis is precisely 58.65 Earth days. The planet rotates precisely three times on its own axis during two full revolutions about the Sun. Mercury's spin-orbit coupling and its rather eccentric orbit lead to some surprising circumstances for a hypothetical inhabitant of Mercury, which can be visualized by reference to Figures 2.3 and 2.4. First, the "day" on Mercury is 176 Earth days long, because the apparent motion of the Sun across the mercurian "sky" repeats every two mercurian years of 88 Earth days' duration. Second, because of Mercury's great orbital eccentricity, the Sun's apparent motion is far from uniform; it even reverses itself briefly! The most important consequence of this orbital eccentricity, however, is the unequal heating of different quadrants of the planet. Longitudes 0° and 180° face the Sun only at Mercury's closest approach to the Sun (termed *perihelion*), whereas 90° and 270° are illuminated only in the vicinity of *aphelion* and receive only about half as much total solar energy.

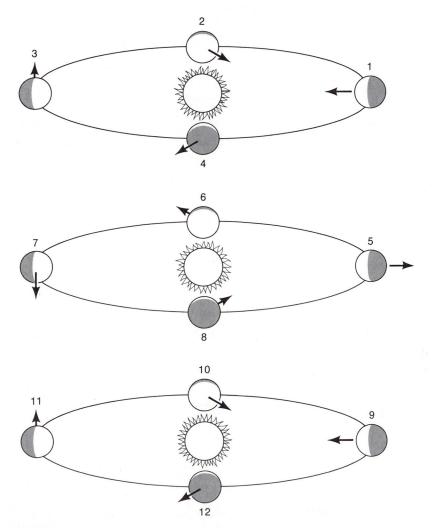

Figure 2.3
The mercurian "day." Mercury rotates three times around its spin axis during every two revolutions about the Sun. In this diagram of Mercury's orbit, the fixed arrow originates from one of the planet's two hot subsolar points—that is, the points on the equator that lie directly under the Sun at alternate perihelions (time of closest approach to the Sun). The sequential numbers reference the position of the planet in its orbit during two of its revolutions around the Sun.

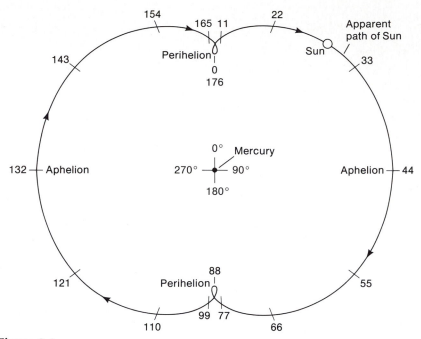

Figure 2.4
The Sun as seen from Mercury appears to execute a loop in the sky during perihelion passage. The apparent position of the Sun in relation to subsolar longitudes on the planet is marked off in 11-day intervals for two mercurian years. [After Steven Soter and Juris Ulrichs, "Rotation and Heating of the Planet Mercury,"*Nature,* **214,** Fig. 1, p. 1315, June 24, 1967.]

The Core of Mercury

It is not only Mercury's unusual rotation that sets it apart from other objects in the Solar System. The planet's density is nearly twice what would be expected for a planet with a Moonlike composition. Mercury is much denser than either the Moon or Mars. Indeed, if the self-compression of a planetary body by its own gravity is allowed for, it can be inferred that Mercury contains a higher proportion of iron than even Earth. Earth's great density reflects the existence of a substantial iron core. Figure 6.1 illustrates the relative sizes of iron core and silicate mantle of Earth and Mercury if both were differentiated more or less as the Earth is. Mercury appears to have a core the size of the

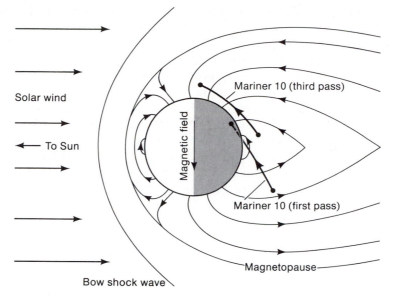

Solar wind

← To Sun

Bow shock wave

Magnetic field

Mariner 10 (third pass)

Mariner 10 (first pass)

Magnetopause

Figure 2.5

The magnetic field of Mercury, unexpectedly detected in Mariner 10's first encounter, was fully confirmed in the third encounter, when the spacecraft was targeted to pass close to the north pole of the planet. The black dots indicate, in planar projection, where Mariner 10 entered and left the planet's magnetic field during each pass. Magnetic field lines are distorted by action of electrically charged particles emanating from the Sun. The north/south polarity of Mercury's field is oriented like that of Earth.

entire Moon. Indeed, the planet can be regarded as mostly an iron sphere enclosed by a relatively thin shell of silicate mantle and crust, especially because Mariner 10 discovered in 1974 a mercurian magnetic field similar in form to that of the Earth, but of much lower intensity (Fig. 2.5). In contrast, the Moon and Venus are both without detectable planetary magnetic fields, and Mars' is barely detectable at best. A large iron core seems required to explain the mercurian magnetic field, although there is less agreement on whether a partly fluid core, as on Earth, is required as well.

A Moonlike Surface

Mercury is a planet of unexpected contrasts. Although its core apparently is Earthlike, its surface is extraordinarily Moonlike.

The mercurian terrain implies that both the sequence of events and the processes were very similar to those that fashioned the lunar landscape.

Before the Mariner 10 mission, Mercury was little more than a fuzzy blur. In March 1974, Mariner 10 flew past the planet and returned pictures showing for the first time details of its surface (Fig. 1.6). Later that same year, because of intentional maneuvers of the spacecraft, Mariner 10 crossed Mercury's path again, allowing a second encounter with the planet and acquisition of additional pictures (Fig. 2.6). Still another encounter was made possible when the spacecraft was manipulated to intersect the same point of the planet's orbit for a third time. During that pass, additional high-resolution pictures were taken and vital diagnostic magnetic-field data were obtained.

Altogether, Mariner 10's three passes provided photographic coverage of about half the surface of Mercury at a resolution comparable to that of Mariner 9 (1971–1972) for Mars and similar to that available for the Moon from Earth-based telescopic observations. Even with only half the surface viewed, the distribution of major physiographic provinces appears to be asymmetric (as on Earth, Moon, and Mars). The side of the planet photographed during the approach to the first encounter showed a heavily cratered surface resembling the lunar highlands (Fig. 1.3). The opposite side of the planet, photographed during the outgoing phase of the flyby, showed both cratered terrain and extensive plains reminiscent of the lunar maria. Also observed were numerous randomly distributed, multi-ringed basins, the largest of which is the 2000-kilometer-diameter Caloris Basin (Fig. 6.5), located near one of the "hot" poles of the planet.

Because Mercury's gravity is higher than that of the Moon (its surface gravity is one-third that of Earth, comparable to that of Mars, and about twice that of the Moon), the *secondary* impact craters, caused by large fragments that fall back to the surface after being ejected during the primary impact, are closer to the main crater rim, giving the cratered terrain on Mercury a slightly different appearance from that on the Moon.

Although some smooth plains photographed by Mariner 10 are reminiscent of those on the Moon, they lack well-defined

Figure 2.6
Orthographic photomosaic of the southern hemisphere of Mercury (centered at 55°S, 100°W) created from images by Mariner 10. With its high frequency of large impact craters, the southern hemisphere closely resembles the terrae areas of the Moon. Both planets lack atmospheres and the associated agents of surface erosion, such as flowing water and wind.

flow fronts and flow channels of the basaltic maria. The question of the origin of many of the smooth plains on Mercury remains unresolved. It has been suggested that they are either volcanic plains, like the lunar maria, or are unusual, fluidized products of the formation of huge impact basins. Detailed mapping shows that there are several different types of plains units, including the oldest unit observed on Mercury—the intercrater plains, upon which many craters and the multiringed basins have been superimposed. The youngest units appear to be plains-forming material superposed on ejecta from the multiringed basins.

One of the most interesting discoveries on Mercury is evidence suggesting thrust-faulting, in which portions of the surface have been compressed, fractured, and one portion forced horizontally over an adjacent portion. This evidence takes the form of arcuate and linear scarps, some of which transect craters and demonstrate compression (Fig. 4.39). Such compressional structures have not been observed on either the Moon or Mars and may be ancient manifestations of the interaction of Mercury's large core with its surface.

Thus the Mercury revealed by Mariner 10 is unexpectedly Moonlike on the outside and unexpectedly Earthlike on the inside. Unraveling those paradoxical results promises new insights into the early bombardment history of the Solar System, and perhaps even into the origin of the Earth's magnetic field.

MARS

For centuries, Mars has intrigued us. The red planet was believed to be the one most likely to harbor life. It was known 200 years ago to have an atmosphere. For nearly a century it was thought to have "canals."

Changing Impressions of Mars

The first close look at Mars was provided by the Mariner 4 flyby in 1965. Its TV pictures showed a cratered surface, suggesting that Mars was simply a larger version of the Moon, in contrast to widespread expectation that it would more closely resemble the Earth. This impression was not substantially changed by two additional flybys, Mariners 6 and 7, in 1969. Their pictures

(Fig. 2.7) also showed a surface littered with craters. There were, however, a number of tantalizing new topographic forms that appeared unrelated to cratering. These included terrains bearing evidence of erosion as well as unusual landforms in the polar regions. Because Mars does have an atmosphere, although a very thin one, strong winds can be generated to erode surface features, transport material, and even cause global dust storms. Moreover, individual craters show differences in morphology from their lunar counterparts. Large martian craters are shallower and generally lack the fresh, bright streaks (rays) that radiate from some similar-sized lunar craters. The absence of such streaks is consistent with wind modification of the martian surface.

The Mariner 9 mission in 1971–1972 radically altered the impression of Mars as a Moonlike object (Fig. 2.8). A unique planetary surface was recognized, fashioned not only by processes like those on the Moon and Earth, but also in new ways not previously encountered. Mariner 9 was an orbiting spacecraft that took pictures of the planet's entire surface at about 1-kilometer resolution. High-resolution frames, revealing detail down to about 100 meters for selected regions, were also acquired for a small portion of the landscape. In 1976–1980, the Viking orbiters greatly expanded the amount, quality, and continuity of high-resolution coverage. Mariner 9 reached Mars during a major storm, which had enveloped the entire planet with a thick cloud of dust that obscured the surface. Thus the process of *eolian* (wind) activity was demonstrated very dramatically. As the dust cleared, a completely new picture of the planet emerged. First to be seen were enormous positive topographic features, known now to be huge volcanoes—much larger than the largest volcanoes on Earth (Figs. 2.9, 4.14, 4.30). They cap an enormous topographic bulge covering nearly a third of the planet's circumference. Mars exhibits greater topographic relief than even Earth.

As the pictures for the entire planet were pieced together, it was seen that, as on the Moon, Mercury, and Earth, the distribution of the major physiographic provinces on Mars is asymmetric. In general, the southern hemisphere of the planet is

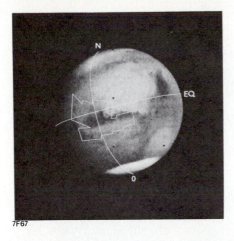

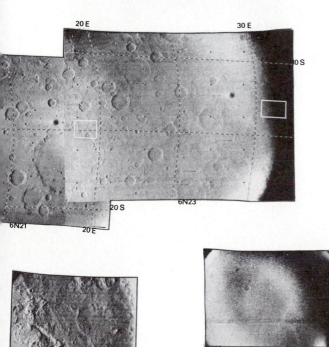

Figure 2.7
Sequence of wide-angle images of Mars obtained during the Mariner 6 flyby in 1969. The inset of the full planet (top right) shows the location of the mosaic; white boxes on the mosaic locate the high-resolution images shown individually around the mosaic. This and similar views of the same general region of Mars—obtained by Mariner 7 and earlier by Mariner 4—led many planetologists to view Mars (incorrectly) as being similar to the Moon in surface history.

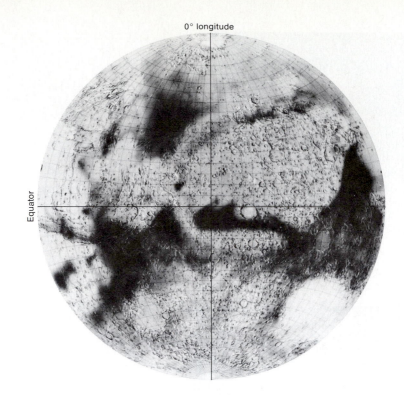

Figure 2.8
The diversity of surface features on Mars was revealed for the first time with the return of images from the Mariner 9 mission. Mariner 9 was placed in orbit around Mars in 1971 and successfully photographed nearly the entire surface of the planet. Shown here are shaded relief maps derived from the Mariner 9 images: Dark areas show low-albedo zones. [Courtesy of J. Inge, U. S. Geological Survey and Lowell Observatory.]

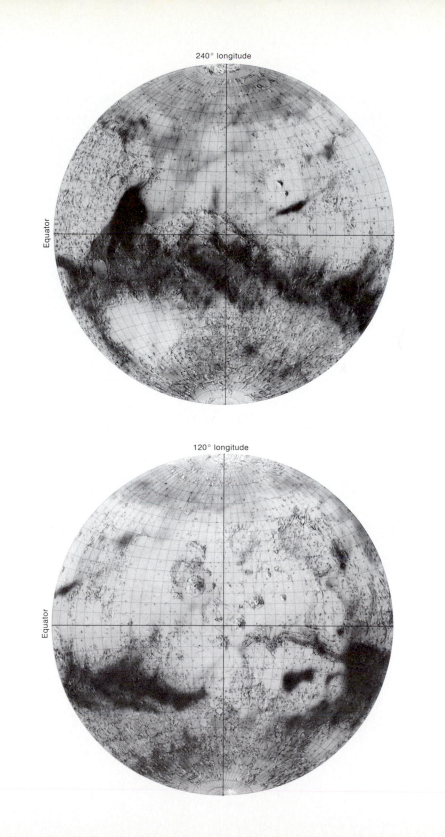

240° longitude

Equator

120° longitude

Equator

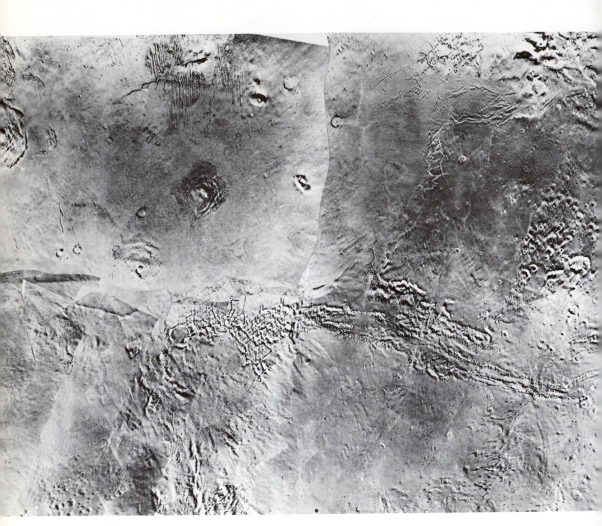

Figure 2.9

Mosaic of Mariner 9 images showing Valles Marineris and the four largest shield volcanoes of the Tharsis area (left side). Valles Marineris stretches several thousand kilometers along the equatorial region of Mars. Although its origin remains enigmatic, even with the abundant high-resolution Viking images that cover it, the western part (left side in this view) appears to be structurally controlled, as indicated by the down-dropped fault blocks. Toward the east, the canyon system grades into the jumbled "chaotic terrain." Illumination for the series of pictures shown here is from the lower left corner.

heavily cratered and includes large circular basins, in many respects similar to the lunar highlands, and is systematically of higher elevation than the northern hemisphere. Much of the northern hemisphere consists of smooth, relatively uncratered plains. As fortune would have it, Mariners 4, 6, and 7 had photographed only southern-hemisphere localities, and thus revealed mainly cratered terrain.

Stretching more than 4000 kilometers along the martian equator is an enormous canyon system named Valles Marineris (Fig. 2.9), which is more than 4 kilometers deep in some places. By comparison, the Grand Canyon of Arizona is a minor feature. The western section of Valles Marineris is dominated by a network of grabenlike features. The valley is widest in the central section, and abundant evidence of landslides can be recognized. Farther east lies a peculiar landscape called chaotic terrain (because of its jumbled and broken surface). This surface appears to feed channel-like structures; in outline, these features resemble some river valleys on Earth, except that they are much larger. It has come to be recognized that at one time there were sufficient quantities of liquid water on Mars to carve these huge channels, probably through catastrophic flooding. Whatever the details of their origin, *fluvial* (flowing water) processes operated on a gigantic scale to form them.

Some of the smooth plains on Mars, when imaged at high resolution, reveal mare-type ridges reminiscent of those on the Moon (Fig. 4.23). These and other topographic features suggest that at least some of the smooth martian plains consist of vast basalt flows. Distinct flow fronts, quite similar to those of the lunar Mare Imbrium, point to a volcanic origin. Indeed, in 1976, Viking 1 landed on what is probably an eroded volcanic surface. Other smooth plains, however, probably consist of vast sheets of consolidated dust that mantle the underlying surface structures. Some craters appear to be draped with a sedimentary deposit, possibly dust. Similarly mantled units are observed in the polar regions.

The surface of Mars, as revealed by the Mariner 9 and Viking orbiter pictures, is extremely complex, even though a basic Moonlike ancient architecture is still preserved in many places. Gigantic and sometimes unique surface features have been created by vigorous and continuing internal activity, as well as by atmospheric modification on a scale unimagined before the 1970s.

The Atmosphere of Mars

The atmosphere of Mars is very tenuous by earthly standards. The surface pressure there is more rarified than at an elevation of 100,000 feet on Earth (i.e., less than 1/100 of Earth's surface pressure). Yet its interaction with the martian surface over several billions of years of geologic time has produced a profoundly different character than that seen on the Moon, Mercury, or even Earth. Indeed, Mars' surface may still record in a few places atmospheric interactions with the surface dating from 4 billion years ago. In contrast, the much denser and more reactive atmosphere of Earth has modified and erased the surface record many, many times over in its history.

Mars' atmosphere is mainly carbon dioxide, along with a few percent of nitrogen and argon. Indeed, a vertical column of martian atmosphere contains about thirty times more carbon dioxide than an equal column on the Earth, where carbon dioxide constitutes only about 0.04 percent of the atmosphere. Water vapor is present in trace amounts in the martian atmosphere and, under some circumstances, forms clouds and ground frosts. However, since the atmosphere is rarely warmer than $-70°C$, the amount of water vapor that can be maintained in such a cold atmosphere is extremely small. Futhermore, because the atmosphere is so rarified, liquid water is unstable and would evaporate or freeze on the surface.

On the other hand, even such a rarified atmosphere can be quite effective in transporting materials along the surface by saltation (jumping) as well as by raising fine material and keeping it in suspension to form enormous dust clouds. Many conspicuous features of the martian surface undoubtedly reflect the operation of eolian processes of this type over millions, perhaps even billions, of years.

The atmosphere is sufficiently thin that the surface temperature in the regions exposed to sunlight is determined mainly by the absorption and reradiation of sunlight, as on the Moon and Mercury. The martian atmosphere has only modest influence on daytime temperatures. By contrast, the Earth's atmosphere greatly modifies and moderates our climate. So do our oceans, by storing and transporting heat.

The Frost Caps of Mars

Since Mars is 1.4 times the distance of Earth from the Sun, its surface is much colder than that of the Earth. Even at noon on the equator, temperatures rarely rise above the freezing point of water. Temperatures rapidly fall with the onset of nighttime, and would rival those of the lunar nighttime except for a unique consequence of Mars' atmospheric composition. On Mars, atmospheric carbon dioxide will freeze to form frost at about $-123°C$, stabilizing the surface temperature at that point. Carbon dioxide frost slowly continues accumulating until daytime solar heating resumes. Nighttime frost tends to form during Winter, early Spring, and late Fall. Mars has seasons very similar to the Earth's because its obliquity (the angle between its spin axis and a perpendicular to its orbital plane) is 24°, nearly the same as that of the Earth. In the late Fall, more carbon dioxide frost forms each night than sublimes during the short days. Thus frost caps of carbon dioxide accumulate alternately at the poles, as Winter comes first to one and then to the other, and disappear by the end of Spring. These annual frost caps, long believed to be composed of frozen water on the basis of a supposed analogy with Earth, were shown by Mariner 7 in 1969 to be composed instead of very dry carbon dioxide frost (Fig. 2.10).

In addition to the seasonal frost caps, small permanent residual caps persist at both poles throughout the Summer. These caps are believed to be composed mainly of water ice, although carbon dioxide in some solid form may also be included.

What would happen if substantial amounts of new carbon dioxide were released at the surface of Mars? The atmospheric pressure probably would not build up much, if at all. Instead, the excess carbon dioxide would condense into frost and accumulate as permanent solid deposits in low areas near the poles (or possibly be adsorbed by the surfaces of the fine particles making up the martian "soil"). The reason is that the present atmosphere is nearly or actually in (annual) pressure equilibrium with condensed carbon dioxide at these two sites. Indeed, a permanent excess of solid carbon dioxide was once proposed to exist near the north pole of Mars. However, measurements

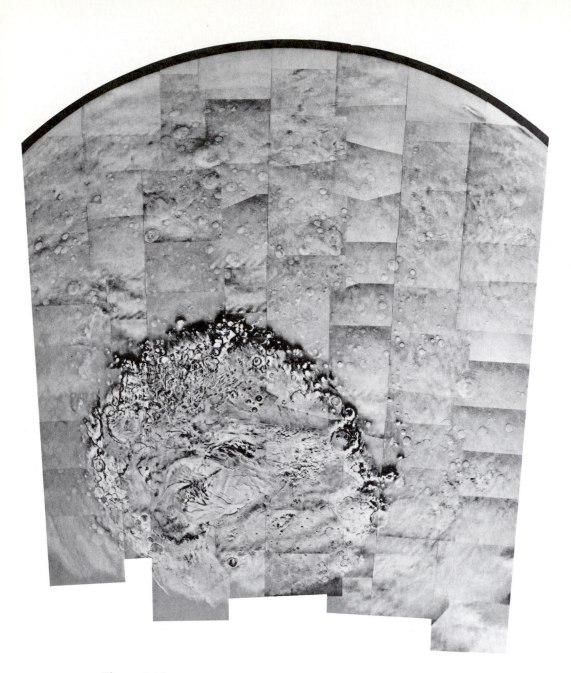

Figure 2.10
Viking orbiter mosaic of the south polar cap of Mars, imaged through a violet filter. Pictures of the frost cap have been enhanced to show detail, with the result that adjacent areas not covered by frost are very dark. Note that near the edges of the cap, ice occurs primarily in the protected areas within craters; "pitted" terrain is visible as gray areas in several parts of the frost cap.

obtained by the Viking orbiter infrared thermal mapping instrument suggest that during the martian summer, the surface temperatures are too high around the north pole to permit permanent carbon dioxide ice to exist. (The situation at the south pole is not so clear.) Thus the carbon dioxide in the ice cap is apparently only seasonal. The existence of some kind of "reservoir" of carbon dioxide (either in solid form or as gas adsorbed by soil grains) has yet to be proved, although its presence is believed responsible for subtle variations in seasonal atmospheric pressure changes.

In any case, the total amount of solid volatiles present at both poles, even if substantial amounts of excess solid carbon dioxide were to exist permanently, is a minute fraction of the abundance of volatiles on Earth. The amounts that can be stored as solids in weathering products or as gases adsorbed by individual soil grains are much larger but still only a fraction of the Earth's abundances. Mars apparently never had a hydrosphere and atmosphere comparable to Earth's, although geochemical and morphological evidence suggest it once had a much more appreciable atmosphere than it does today.

Climatic Fluctuations on Mars and Earth

In the polar regions, there are characteristic wind-sculptured cliffs composed of evenly layered units (Fig. 3.27) ranging tens of meters in thickness. It has been suggested that these layers were deposited as mixtures of wind-transported dust and precipitated ice and have subsequently been eroded to form these beautiful, uniquely martian, terraced cliffs.

Whatever the details of their origin, the uniform character of these widespread layered deposits discovered by Mariner 9 at both martian poles strongly suggests that global atmospheric conditions fluctuated on a time scale affecting the formation of deposits tens to hundreds of meters thick. A mechanism that likely triggered such atmospheric changes on Mars has been identified; the obliquity of Mars is not constant at its present value of 24° but is now recognized to fluctuate between limits as high as 35° and as low as 15° because of perturbations from other planets, especially Jupiter. This is a much larger change in obliquity than for the Earth.

As illustrated in Figure 2.11, changes in the obliquity of Mars must have a strong effect on the temperatures of the polar regions and thus on the formation of seasonal and permanent frost deposits. When the obliquity of Mars decreases from 24° to the minimum of 15°, the average annual temperature of the polar frost deposits decreases, causing even more carbon dioxide to freeze, so that the vapor pressure of carbon dioxide at the surface is reduced by as much as a factor of ten. When the obliquity exceeds 24°, the process reverses, causing any excess solid carbon dioxide on or just beneath the surface to vaporize. If there is in fact an excess of carbon dioxide in the solid form now, an increase in obliquity could raise the surface pressure by as much as five or six times the present value. Such variations in obliquity take place periodically over time scales lasting hundreds of thousands and millions of years. Very likely the climate of Mars has fluctuated throughout at least the past few billion years of the planet's history.

Earth also has experienced fluctuations in polar temperatures a few times in the past billion years. The most recent period of fluctuations was responsible for the glacial cycles of the Pleistocene Epoch. Indeed, one controversial theory of glacial cycles advanced long ago ascribes a major cause of glaciation on Earth to the small fluctuations in Earth's spin orientation and orbit, which, like those of Mars, also arise from tiny but periodic gravitational effects from Jupiter and other planets.

Figure 2.11
This figure illustrates periodic variations in Mars' orbital eccentricity and the angle between its spin axis and the Sun (its obliquity) caused by the gravitational influence of the other planets, primarily Jupiter. Also shown are the resultant variations in the amount of sunlight reaching the martian surface (insolation). The top plot shows that the eccentricity of the martian orbit has varied from a nearly circular 0.009 to a relatively elliptical 0.14. The second plot shows that heating of the subsolar spot on the surface at closest approach to the Sun (perihelion) closely mimics the eccentricity variations. The periodicities of these variations are 95 thousand years for the high-frequency changes and two million years for the low-frequency changes. Obliquity (middle plot) also varies periodically, at times reaching a low of 15° and at other times a high of 35°. These variations occur on a time scale of 120,000 years and 1,200,000 years. Variations in sunlight reaching the polar regions are illustrated in the bottom two plots. Both the amount of light reaching a given pole and the difference between one pole and the other show periodicities of about 55,000 years within a broader cycle of about 1,500,000 years. These variations in insolation most likely affect the climate of Mars, certainly reducing its temperature and atmospheric pressure at some times and perhaps increasing them at other times.

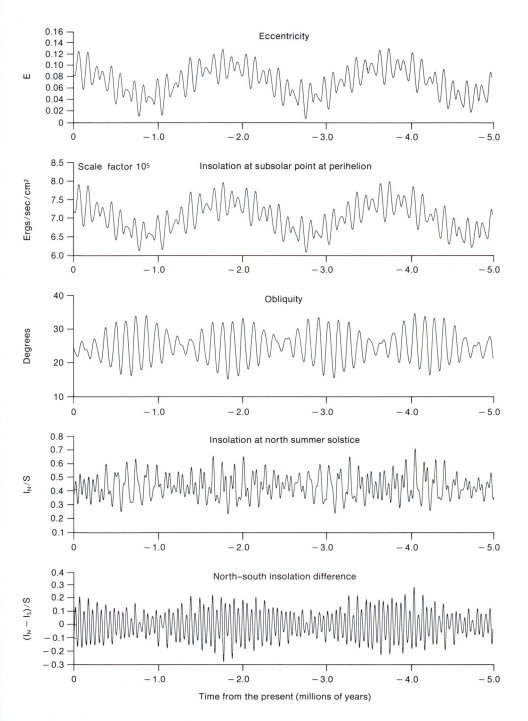

But the Earth's heat balance is far more complex than Mars' because of the role played by oceans. A fascinating question about the two planets is whether the causes of the ice ages on Earth and of the layered deposits on Mars are related. Were they both produced by external causes, corresponding perhaps to a decrease in energy received from the Sun? Or are Earth's ice ages caused and modulated by internal factors, such as variations in global volcanism? If there has been any change in solar illumination over the past 3 million years, the martian climate must also have been affected in a characteristic way. A record of any such changes may exist in the polar regions of Mars in the form of fine-scale bedding within the layered polar deposits.

Perhaps martian deposits preserve records of even shorter-term fluctuations in solar intensity, a subject of great importance to food production on our crowded Earth. Just as we may discover otherwise unobtainable facts about the Earth's magnetic field through study of the magnetic field of Mercury, so we may be able to isolate the primary causes of glaciation and climatic change on Earth by studying the "control" planet— Mars. It has no oceans to complicate matters, and it is still heated by the same Sun.

What about longer-term climatic change on Mars? Could a denser, aqueous atmosphere have existed several billion years ago to spawn the surface fluids that created some of Mars' sinuous channels? Its surface environment seems to have been quite different billions of years ago from the one that exists today. Probably the atmospheric pressure was higher and other volatiles besides carbon dioxide were abundant billions of years ago. A genuine geological prize lies encrypted within the strange and bewildering terrains of Mars: the origin and evolution of the martian atmosphere and their implications, if any, for the development of our own atmosphere, hydrosphere, and biosphere.

Life on Mars?

The idea of a habitable Mars developed strongly in the minds of some scientists and popular writers because of its presumed similarity to the Earth. Some observers even thought they saw "canals" on Mars through telescopes. Mars exhibits nearly the same rotation rate and same obliquity as the Earth, seasonal changes in frost caps, and also a reasonably consistent pattern of seasonal changes in light and dark markings. Hence the notion that Mars is really Earth's twin was fostered by the erroneous conclusions that the present similarity in spin reflects similarities in origin, that the frost caps are evidence of a water-rich surface environment, and that changes in light and dark markings are the consequences of the seasonal development of plant life. In fact, the present similarity in obliquity is purely fortuitous, since the obliquity of Mars varies over a range of 20°, whereas that of the Earth is nearly constant. The present similarity in spin rate is likewise fortuitous, since the Earth's rotation rate has been gradually slowed by the presence of the Moon while Mars' has remained constant. The martian frost caps consist of frozen carbon dioxide, not frozen water; Mars is far dryer and more hostile to liquid water than was ever imagined before Mariner 4 first flew by Mars in 1965. Seasonal dust storms, not plant life, are the causes of the variations in markings observed through Earth's telescopes.

Why, then, has there been such enthusiasm for the idea of life on Mars if there is no evidence supporting that conclusion? Mars probably never manifested any primitive oceanic environments like those hypothesized on Earth, in which rich concentrations of organic chemicals could accumulate and interact, presaging the momentous accident when the first self-replicating molecule was created by chance. The very large old craters on Mars, which may have formed 4 billion years ago (as on the Moon and perhaps Mercury), would have been eroded beyond recognition in the aqueous environment necessarily created by the existence of oceans.

Thus, if life were discovered on Mars, it would mean either that life could form in another way than is generally believed to have occurred on Earth or that our current views of the origin of life on Earth may be incorrect. On the other hand, the absence of evidence of microbial life, as suggested by the results of the U.S. Viking landers in 1976, is merely consistent with most scientific expectation. The possibility of alien life is so significant that the question must be asked scientifically each time a possible new habitat is encountered.

VENUS

The Densest Atmosphere

Less is known about Venus than any of the other Earthlike planets. It has an extremely dense, cloudy atmosphere that completely obscures its surface in visible light. Through Earth-based optical telescopes, Venus appears as a very bright, but completely featureless, object. The nature of the particles that compose its bright cloud layers has baffled scientists for many years. Recently there has been growing evidence that these particles are, at least in part, droplets of *sulfuric acid!* In addition, Venus has trace amounts (in gaseous form) of two other especially corrosive acids, hydrogen fluoride and hydrogen chloride, as well as sulfur dioxide in its lower portions. As on Mars, the principal atmospheric constituent is carbon dioxide. Water vapor is present at about 0.1 percent.

Even more surprising than the composition of the upper part of the atmosphere are the pressure and temperature at the bottom. The surface pressure on Venus is nearly 100 times that on Earth, and the temperature at the surface is around 700° absolute. Venus is uniformly, oppressively hot; surface conditions must be quite insensitive to small changes in solar insolation. Thus the climate of Venus is probably quite stable.

What an unearthly place! Yet planetary scientists regard Venus as the twin of Earth in many ways. As can be seen in Figure 2.12, Venus resembles the Earth in size and mass and therefore probably in internal constitution. Its surface chemistry, as crudely revealed by the gamma-ray spectrometers on board the Venera landers, is not unlike that of common Earth rocks. In addition, the enormous amount of carbon dioxide gas in the atmosphere of Venus is comparable to the total amount of carbon dioxide that has been in Earth's atmosphere over geologic time. However, nearly all of Earth's carbon dioxide recycles through various states. The largest fraction presently resides in the upper crust, primarily in the form of solid carbonate minerals that compose the limestones and dolomites. Earth's oceans, where these rocks were formed and continue to be formed, today contain a substantial reservoir of dissolved

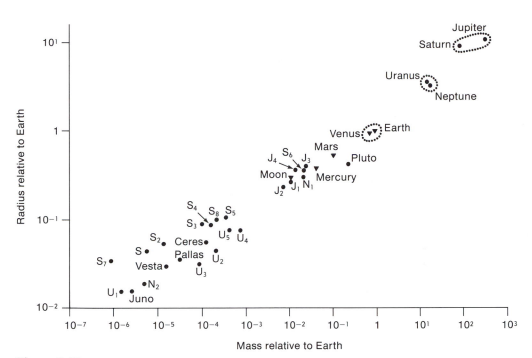

Figure 2.12
Diagram showing mass versus size for the planets, larger satellites, and larger asteroids in the Solar System. The letters J, S, U, and N refer to satellites of Jupiter, Saturn, Uranus, and Neptune; triangles show the terrestrial planets, including Earth's Moon. The similarity of Venus/Earth, Uranus/Neptune, and Saturn/Jupiter is emphasized in this log/log plot.

carbon dioxide in equilibrium with the gaseous form in the atmosphere. Venus and Earth both have released to their atmospheres comparable amounts of carbon dioxide, but the carbon dioxide on Venus is nearly all still in the gaseous phase in the atmosphere. Surface temperatures on Venus reach extraordinary levels because of the effectiveness of carbon dioxide (along with the minor amount of water vapor) in trapping the heat reradiated by the surface and absorbed by cloud particles.

The principal chemical difference between the Earth and Venus is the absence of water on the surface of Venus. Earth, indeed, is the watery planet, and much of its uniqueness geologically (and all of its uniqueness biologically) can be attributed to that characteristic. Venus may never have had abundant water. Or perhaps Venus originally contained a similar amount of water but lost it through evaporation of the oceans and en-hanced atmospheric breakdown of water molecules.

On Earth, a delicate balance prevails between absorbed sun-light, reradiated heat, and the abundance of heat-retaining gases in the atmosphere, such as carbon dioxide and water vapor. If the concentration of heat-retaining carbon dioxide is increased in the atmosphere (by continued combustion of fossil fuels and by continued deforestation of virgin lands), the sur-face temperatures tend to be increased, leading possibly to fur-ther release of these gases from the oceans into the atmosphere. Thus the process might continue if it is not moderated by other effects. Perhaps this hypothetical condition of excess atmos-pheric carbon dioxide actually prevailed early in Venus' his-tory, destroying primitive oceans and leaving instead an arid, corrosive planetary surface and atmosphere.

The Strangest Motions

Although Venus is featureless when viewed in visible light, photographs taken in the invisible light of shorter wavelengths than blue—the ultraviolet—show faint but recognizable atmos-pheric markings. Such photographs, taken through Earth-based telescopes, reveal that the markings display an approximately four-day retrograde period. However, radar observations of the surface show that the body of the planet rotates much more slowly, once every 243 days. This rotation is also *retrograde*—backwards compared to the rotation of the Sun and almost all the other planets. Thus the purported atmospheric rotation ob-served in ultraviolet photographs must be sixty times *faster* than the planetary spin, certainly not an intuitively plausible circumstance. Nevertheless, in 1974, Mariner 10 photographed Venus repeatedly in the ultraviolet from closeup (Fig. 1.5) and verified not only the existence of an organized pattern of ultra-violet markings, but that the equatorial rate of rotation of the markings is indeed approximately once every four days. Soviet entry-probe data also indicate high velocity at high alti-tude. The results were further confirmed and extended by the Pioneer Venus orbiter and probes in 1979.

The phenomena responsible for these markings take place

high in the atmosphere of Venus and involve poorly understood processes of scattering and/or absorption of ultraviolet wavelengths. The pressure there is perhaps ten percent that on the Earth's surface and one-tenth of a percent that on Venus' surface. Wind velocities are quite high (about 100 meters per second), and even higher velocities can be found toward the poles.

How could such a strange atmospheric circulation system develop? Why is Venus' atmosphere different from Earth's and Mars'? The principal difference is that both Earth and Mars absorb sunlight primarily at the surface. Their atmospheres are heated from the bottom. Strong latitudinal differences in surface temperature are created and, in turn, drive the atmosphere to transport heat from the warmer equatorial regions to the colder polar regions. In addition, both rapidly spinning planets, Earth and Mars, exhibit a strong Coriolis effect, which arises from the interaction of the meridional motion (north-south winds) with the planetary spin and results in three-dimensional patterns of motion.

On Venus, by comparison, a smaller fraction of the incident solar energy penetrates the atmosphere all the way to the surface. Most is scattered back into space to provide the bright image that is seen through the telescope; much of the remainder is absorbed within the atmosphere. Thus Venus' atmosphere is heated more at the top and middle than at the bottom and, in this sense, resembles more the shallow seas of Earth than its atmosphere. Furthermore, Venus' surface rotates very much slower than Earth's or Mars'.

Laboratory experiments showed some years ago—with a much simpler case of heating from the "top" (a rotating annular container heated at one point by a flame)—that a surprising circulation pattern can develop. The fluid can be caused to rotate considerably faster than the relative rate of motion of the point at which heat was applied to the rotating fluid. Indeed, it has been suggested that a three-dimensional version of these idealized experiments—in which the Sun acts as a "moving flame"—might operate on Venus. Data from Mariner 10 and the Pioneer Venus mission support such novel concepts and open the way to further detailed understanding of this unique atmosphere.

Why Does Venus Spin Backwards?

The gradual slowing of the Moon's rotation through tidal inter-action with Earth is a plausible explanation of the present synchronization of the Moon's rotation with that of the Earth. Similarly, tidal interaction with the Sun plausibly could have produced Mercury's 3:2 spin-orbit coupling. Mercury's spin rate probably has evolved over geologic time from a faster, perhaps Earthlike spin rate. Eventual synchrony between its spin rate and orbital period about the Sun may be that planet's destiny.

But what process could have produced *retrograde* rotation on Venus? Certainly it is improbable that Venus simply *began* by spinning opposite to other planetary bodies in the Solar System (and to the direction of spin of the Sun itself). Hence, ever since Earth-based radar discovered the extraordinary spin of the planet Venus in the early 1960s, this phenomenon has constituted a challenging question of the inner Solar System.

One conjecture is that Venus might once have captured a moon in retrograde orbit whose orbit subsequently decayed until that moon crashed into the planet. This conceivably might have started the planet spinning retrograde. Perhaps a more likely possibility is that the atmospheric "moving-flame" effect itself contributes to the peculiar spin. In the laboratory experiments referred to above, it was found that the rotation of the fluid can be in either direction compared to that of the heat source, depending on circumstances at the start of the experiment. Consequently, it is conceivable that the moving-flame effect on Venus initially created retrograde winds that, combined with tidal interaction with the Sun, caused a gradual change from direct (termed prograde) to retrograde rotation.

Another surprising indication of dynamic coupling among the Earthlike planets is the fact that the 243-day retrograde spin of Venus means that when the Earth and Venus are at closest approach, Venus always tends to present the same side toward the Earth! Hence gravitational coupling between the Earth and Venus is suggested, although such a connection is intuitively difficult to accept in view of the great distances and very small gravitational effects involved. In fact, the exact rotation period of Venus appears to differ slightly but significantly from that corresponding to exact Earth-Venus synchronism.

First Glimpses of the Surface

Scant information about the surface of Venus comes from three primary sources: (1) measurements made by U.S. Earth-based radar installations, (2) measurements from successful automated capsules landed there by the Soviet Union, and (3) recent measurements from a moderate-resolution radar system aboard the U.S. Pioneer Venus Orbiter.

Radar can "see" through thick clouds to return information about the underlying solid surface beneath them. In one form the reflected radar signals can be used to derive topographic elevations; in another, they can be used to reconstruct images of the surface that, when combined with the radar topographic data, produce a recognizable portrayal of the planetary surface. Preliminary radar results suggest that Venus' surface topography may very well be every bit as interesting as Mars'.

Since Venus is of about the same size and mass as the Earth—and perhaps has a similar internal structure and thermal history—it may also exhibit the kind of crustal deformation manifested on Earth, termed *plate tectonics*. This mechanism is responsible for the existence of continents and ocean basins, rift valleys, island arcs, and the other principal geographic features on the Earth.

In addition, the extraordinary atmosphere of Venus may well have had a pronounced effect on the surface geology, perhaps quite different in detail from that of either Earth or Mars. The present radar "pictures" of the surface show some features that are tantalizingly similar to what major crustal features of the Earth might look like if observed at very low resolution by radar (Fig. 4.15). Yet others seem to resemble the gigantic circular impact basins of the Moon and Mercury. It seems unlikely that both kinds of features really coexist on Venus unless the surface survives from an early, unique espisode. New spacecraft systems as well as improved ground-based radar observations give hope of providing much improved radar pictures in the future.

Five Soviet and four U.S. spacecraft have landed on the surface of Venus and returned fragmentary data about it. Venera 8, in 1974, obtained preliminary chemical information suggesting a composition similar to Earthlike granite. Venera 9 and 10 landed in late 1975 and returned the first closeup pictures of the surface. (Venera 11 and 12, in 1978, produced new atmospheric data plus evidence of "thunder" and "lightning," but they evidently failed to return data from the surface itself.) Seen from the Venera 9 and 10 spacecraft were surfaces littered with rocks and boulders (Fig. 2.13); some were angular, some rounded, suggesting weathering and active surface processes on the planet. Just as early oceanographers imagined that the deep bottoms of the ocean were extremely smooth, unchanging, and unaffected by active processes, so planetologists— until these pictures were returned—had expected the surface of Venus to be similarly uninteresting. In fact, parts of the deep oceans, when photographed from special spacecraftlike vehicles, also exhibit evidence of unexpectedly rapid erosion and transportation. Apparently Venus' surface is likewise affected by more active surface processes than anticipated. Clearly it is an exciting new planet to explore, perhaps the one with the most information to reveal pertinent to our own planet Earth.

Figure 2.13
Surface of Venus at two sites as viewed by the Soviet Venera 9 (top) and Venera 10 (bottom) spacecraft in 1975, showing rocks and fine-grained deposits. The degraded character of some of the rocks, some of which are about a meter across, records the action of weathering and erosion, but the presence of relatively angular fragments implies that contemporaneous renewal processes are also at work. The white object at the bottom of each image is part of the spacecraft.

SUGGESTED READING

Arvidson, R. E., A. B. Binder, and K. L. Jones. "The Surface of Mars." *Sci. Am.*, **238**:76–89 (March 1978). (Offprint No. 399.)

Dunne, J. A., and E. Burgess. *The Voyage of Mariner 10: Mission to Venus and Mercury.* NASA SP-424. Washington, D.C.: U.S. Government Printing Office, 1978.

Guest, J. E., and R. Greeley. *Geology on the Moon.* London: Wykeham Publications, 1977.

Hartmann, W. K., and O. Raper. *The New Mars: The Discoveries of Mariner 9.* NASA SP-337. Washington, D.C.: U.S. Government Printing Office, 1974.

Hess, W., R. Kovach, P. W. Gast, and G. Simmons, "Exploration of the Moon." *Sci. Am.*, **221**(4):54–72 (October 1969). (Offprint No. 889.)

Horowitz, N. H. "The Search for Life on Mars." *Sci. Am.* **237**(5):52–61 (Nov. 1977). (Offprint No. 389.)

James, J. N. "Voyage of Mariner 2." *Sci. Am.* **209**(1):70–84 (July 1963).

James, J. N. "Voyage of Mariner 4." *Sci. Am.* **214**(3):42–52 (March 1966).

Leighton, R. B. "Photographs from Mariner 4." *Sci. Am.* **214**(4):54–68 (April 1966).

Leovy, C. B. "The Atmosphere of Mars." *Sci. Am.* **237**(1):34–43 (July 1977). (Offprint No. 369.)

Levin, E., D. D. Viele, and L. B. Eldrenkamp. "Lunar Orbiter Missions to the Moon." *Sci. Am.* **219**(5):58–78 (May 1968).

Mariner 9 Television Team. *Mars As Viewed by Mariner 9.* NASA SP-329. Washington, D.C.; U.S. Government Printing Office, 1974.

Mason, B. "Lunar Rocks." *Sci. Am.* **225**(4):48–53 (Oct. 1971).

Murray, B. C. "Mars from Mariner 9." *Sci. Am.* **228**(1):48–69 (Jan. 1973).

Murray, B. C. "Mercury." *Sci. Am.* **233**(3):58–68 (Sept. 1975).

Murray, B. C. and E. Burgess. *Flight to Mercury.* New York: Columbia Univ. Press, 1977.

Mutch, T. A. *Geology of the Moon.* Princeton University Press, 1970.

Mutch, T. A., R. E. Arvidson, J. W. Head III, K. L. Jones, and R. S. Saunders. *The Geology of Mars.* Princeton University Press, 1976.

Pioneer Venus Preliminary Science Reports (1969), Science **203** (no. 4382), 743–808.

Pollack, J. B. "Mars." *Sci. Am.* **233**(3):106–117 (Sept. 1975).

Schultz, P. H. *Moon Morphology.* Austin: Univ. Texas Press, 1976.

Schurmeier, H. M., R. L. Heacock, and A. E. Wolfe. "Ranger Missions to the Moon." *Sci. Am.* **214**(1):52–67 (Jan. 1966).

Sloan, R. K. "Scientific Experiments of Mariner 4." *Sci. Am.* **214**(5):62–75 (May 1966).

Viking Lander Imaging Team. *The Martian Landscape*. NASA SP-425. Washington, D.C.: U.S. Government Printing Office, 1978.

Wood, J. A. "Moon." *Sci. Am.* **233**(3):92–102 (Sept. 1975).

Young, A., and L. Young. "Venus." *Sci. Am.* **233**(3):70–78 (Sept. 1975).

3

MODIFICATION
FROM WITHOUT

MODIFICATION FROM WITHOUT

In expanding the understanding of the surfaces of the Earthlike planets, planetologists must distinguish between the handiwork of competing internal and external processes. Volcanism and deformation of the surface, both the result of internal heat generation, are the principal sources of modification from within. External phenomena that contribute toward sculpturing planetary surfaces include impact cratering, landslides, wind storms, and the action of running water.

PLANETARY SURFACE CYCLES

The surfaces of all the Earthlike planets are still undergoing changes. Part of the Surveyor III spacecraft that landed on the Moon in 1967 was found to be slightly "weathered" when recovered by Apollo astronauts nearly two years later. Mars exhibits dust storms and abundant wind-created deposits like sand dunes. Indeed, Mars' erosional features are so gigantic compared with Earth's that one wonders if there is any balancing of constructive and destructive surface processes as there is on the Earth. The pictures of Venus taken by Venera 9 and 10 indicate that it apparently has freshly exposed material on its surface. Somehow, fresh bedrock must be uncovered there despite favorable conditions for chemical weathering and, presumably, the development of a thick mantle of fine material.

The Surface Cycle of the Earth

The materials that make up the Earth's continental crust generally have been through many chemical cycles. Fresh rock exposed on land (or on the sea floor) is broken up in place by both physical and chemical weathering. In nearly all environments on Earth, the dominant weathering process is chemical: Water can fairly readily dissolve a limestone, and it can, given plenty of time, completely alter the minerals of such tough rocks as granite to clays soft enough to be cut with a pocketknife. This is particularly true in humid areas. Disintegrated and dissolved materials are transported from their original location by gravity, wind, and water (e.g., landslides, dust storms, and rivers). Weathering and transportation together constitute *erosion*. When the transporting medium can no longer maintain sufficient carrying capacity, the transported material is deposited to form a sediment; upon eventual burial and consolidation, sediments become sedimentary rocks. On Earth, most sedimentary rocks are formed within and upon the margins of continents; the deep seas account for only a small fraction. Subsequently, burial, heating, and deformation may transform sediments into recrystallized rocks. When such renewed crystalline rocks are exposed at the surface through uplift and the downcutting of erosion, they are subject anew to the same cycle of erosion and sedimentation.

Thus, within the Earth's continents, many rock constituents have been recycled physically as well as chemically. Chemical interaction with the atmosphere and hydrosphere (which leads to chemical fractionation and differentiation by weathering) is believed to be a significant mechanism for the development of distinct continental and oceanic crusts.

Chemical Recycling on Other Planets?

What about the other planets of the inner Solar System? Have their surface materials also been chemically recycled through the billions of years of their history? The answer is probably "no" for the Moon and Mercury. There are significant differences in chemical composition between the lunar crust underlying the dark plains and that under the brighter highlands. Perhaps the crust of Mercury also shows chemical differences.

However, such chemical difference must have persisted since shortly after the crusts of those planets first cooled and solidified. They are not the result of a continuing geochemical separation like that associated with the evolution and growth of Earth's continents over billions of years.

Mars probably also manifests variations in crustal composition. But so far it does not appear that surface processes have contributed significantly to chemical segregation of the crust. The Viking Lander soil analyses indicate a high calcium-to-potassium ratio. This suggests that, at least for the Viking 1 landing site, the crust does not contain a large fraction of granitic or other similar rocks. These results are consistent with the earlier photogeologic interpretation of the region as being underlain by basaltic plains in which granitic rocks (i.e., rocks with a high silica content) would not be expected. On the other hand, the surface chemical analyses suggest that soluble salts are, or once were, present on Mars. Such salts conceivably could be the residuum of an earlier, more aqueous epoch, but more likely are related to the extensive and more recent volcanism there. The detailed chemical history of Mars' surface is a scientific question of paramount importance that can be answered satisfactorily only after numerous and detailed chemical analyses of surface materials become available, probably through automated return of samples to Earth.

Venus, the Earth's twin in so many ways, likewise could have a tectonically active crust, perhaps even including chemical recycling analogous to the situation here on Earth. Very limited chemical data acquired about the surface by the Soviet Venera spacecraft suggest to the Soviet scientists that chemically differentiated crustal material is present. However, much more comprehensive analyses must be made before we will know whether those first clues really are significant. Thus the chemical history of Venus' surface remains a key question.

Surface Processes on Bodies Without Atmospheres

What are the effects of the differing surface environments of the Earthlike planets on the characteristic surface processes that operate there? Neither the Moon nor Mercury has an atmosphere now, nor did either planet ever have a permanent atmosphere. There cannot have been any chemical weathering from interaction with water vapor, oxygen, or carbon dioxide, nor could transportation or sedimentation have been accomplished by either wind or water.

Nevertheless, some surface processes are greatly enhanced on those planets. For example, impact cratering is much more effective on airless bodies, because even tiny objects can penetrate to their surfaces and the debris resulting from impact is sprayed over much larger distances. Indeed, impact is the principal process that has influenced the surface character of both the Moon and Mercury.

Moreover, all wavelengths of solar radiation reach the surface of planets without atmospheres, including very energetic invisible ultraviolet radiation, which, on Earth, is generally absorbed by the ozone layer of our upper atmosphere. Ultraviolet radiation can induce changes in the optical properties of minerals, and can even cause crystal structures to be broken apart and microscopic chemical separation (sputtering) to take place. In fact, these very effects have been invoked by some to explain the apparent darkening with time of fresh materials exposed by impact on the surface of the Moon and Mercury. It has long been recognized that fresh impact craters generally display radial ray patterns that are brighter than their surroundings; older impact structures do not display these features as conspicuously, indicating the action of some aging process. After examination of the lunar soil samples brought back by the Apollo astronauts, we now know that the impact process also produces (by melting) large amounts of glass beads whose optical properties differ from those of crystalline material of the same composition. Because these glass beads may tend to devitrify with time and become buried by further cratering, it has been argued that they may account for much of the initial brightness of the ray patterns and of their subsequent apparent darkening.

Mars and the Antarctic Connection

The atmosphere of Mars is so rarified—the equivalent of the Earth's at 100,000 feet or higher—that it shields little of the Sun's ultraviolet rays. As a result, water vapor dissociates into hydrogen and oxygen, and oxygen atoms further combine to form ozone—right at the surface. It has been known for some time that laboratory exposure of common mineral powders to simulated martian irradiation and atmospheric conditions can lead to chemical alteration. Just how important a role such "unearthly" surface chemical reactions play in weathering and soil formation on Mars was a principal discovery of the Viking Landers. Oxidation is far more complete there than anywhere on Earth, probably including the formation of peroxide and superoxide compounds in the soil. As a result, any existing organic material would be very rapidly oxidized into gaseous carbon dioxide. Very sensitive measurements made by the gas chromatograph/mass spectrometer on the Viking Landers failed to detect any organic compounds at all! By comparison, the same instruments, prior to the Viking flight, were able to detect components of organic compounds in all terrestrial soils tested, even those from the most inhospitable environments known, such as the dry valleys of Antarctica.

On Earth, there is no exact analog to the martian environment. Probably the closest resemblance is to be found in the dry valleys of Antarctica (Fig. 3.1), where temperatures are so low and precipitation so light that fallen snow generally sublimes directly to vapor rather than melting. These dry valleys have been studied for many years, most recently as possible martian analogs. Summer temperatures range from about 24°C to barely above the freezing point of fresh water. There are estimated to be 20–60 freeze/thaw cycles per year in the upper few feet of the surface soil above the permanently frozen ground. Unlike Mars, these dry valleys even receive some meltwater runoff from nearby glaciers near noontime in summer. All runoff is confined to the valleys, where it drains into saline lakes. Soluble salts are found in the soils and on the surface throughout, although their origin still is not fully established. As on Mars, volcanism may be the source, but the peculiar weathering circumstances there cannot be entirely excluded.

Like Mars, the region experiences high winds, with velocities of 150 to 200 kilometers per hour. The principal effect of chemical weathering is oxidation of the ferrous iron in minerals to form hydrated iron oxides (limonite); a slightly orange color can result. Of course, no ozone is present, as there is on Mars. Physical weathering takes place very effectively, especially through repeated freeze/thaw action.

Figure 3.1
Unusual weathering forms in Antarctica are shown in this aerial photograph of the labyrinth region of the Wright Dry Valley, Antarctica. Upper Wright Glacier is in the lower left and McMurdo Sound in the distance. [U.S. Navy photograph.]

Of possible interest in relation to Mars is a controversial interpretation of the effect of salt on weathering. If salt is so abundant on a surface that the small amount of moisture held in the pores and cracks of rocks becomes saturated, the growth of salt crystals in such pores and cracks could serve as an effective process of disintegration, perhaps producing a large region of peculiar topography. Such a process has been suggested for Mars, where salts may be abundant on the surface. Hence, even though the martian atmosphere is too cold and too rarified for fresh water to exist on the surface, soil moisture may exist on Mars through saline freezing-point reduction.

In terrestrial Arctic regions, where liquid water is more commonly present now than on Mars or in the Antarctic dry valleys, ice can become incorporated in the surface materials and compose up to 90 percent by weight of the "rock." Exposure of such ice-laden materials due to downcutting by rivers or wind erosion can lead to the ablation of the ice and subsequent collapse and erosion of the deposit. The resulting terrain exhibits topographic features similar to those seen in certain areas of Mars (Fig. 3.2). Whether Mars actually has experienced erosional conditions comparable to those of this Arctic analog or whether salt weathering is significant there is uncertain.

There has been an enormous amount of erosion on Mars, especially in equatorial regions. Just where the eroded material went is a major problem. This is an important question in attempting to understand the cycles and processes operating on the martian surface. In the past, surface conditions on Mars may have been dramatically different, probably including an early phase with a moderately dense atmosphere. It is important to remember that some martian surface features may be due to processes for which there are simply no terrestrial analogies.

The Unearthly Surface of Venus

The closest thing on Earth to the surface environment of Venus might be the mouth of a volcanic fumerole at a depth of 1 kilometer beneath the ocean! The fluid pressure, the ambient temperature, and the presence of corrosive gases form the basis of

Figure 3.2
"Fretted" terrain on Mars. Photomosaic of Mars' Nilosyrtis region. Centered at about 34°N, 290°W, this 150-by-80-kilometer (93-by-50-mile) area is a transitional zone between an ancient cratered terrain to the south (right) and sparsely cratered terrain to the north. Principal features in the low-lying areas are suggestive of flow. They resemble some Earth features where near-surface materials flow en masse very slowly, aided by the freeze/thaw of interstitial ice—water frozen between layers of ground materials.

such a surprising comparison. In any case, there can be little doubt that chemical weathering must be very effective on Venus' surface. Elements such as sulfur, phosphorus, and chlorine are abundant there, and should tend to react with rock-forming silicon, iron, and magnesium to form stable solids and gases. In contrast to Mars, however, it is hard to delineate the role of physical weathering and wind transport on Venus. The daily fluctuations in temperature at the surface must be minimal. Most theoretical considerations of the atmosphere, as well as the few relevant observations, suggest that wind velocities close to the surface are very low. Nevertheless, the pictures from Venera 9 and 10 do show, in some places, fresh-appearing rock fragments, indicating that renewal processes of some kind must indeed be operating on the surface of Venus.

The biggest uncertainty is the role of water, both present and past. In the observable part of the upper atmosphere, water exists only as a gas and only in trace amounts; in the deeper atmosphere or on the surface, it can exist only as steam. Indeed, the paucity of surface water may well be the principal difference between Venus and the Earth. Whether water ever was abundant on Venus' surface is one of the major unanswered questions about the planet.

UBIQUITOUS IMPACT—FROM MICROSCOPIC CRATERS TO THOUSAND-KILOMETER BASINS

Craters catch the eye as the dominant features on most planetary surfaces (Fig. 3.3). Ranging in size from the enormous basins that form the "eyes" of the "lunar face" down to the limits of telescopic resolution, lunar craters have fascinated moon-watchers for more than 300 years. Most plausible explanations held either that they were of volcanic origin or were caused by impacting objects. It was not until the great expansion of geologic studies of the Moon during the 1960s that sufficient evidence was gathered and criteria were well enough defined to permit their confident identification as impact craters. We now know that the great majority of the craters on the Moon and on Mercury, and many on Mars as well, were caused by impacts.

MODIFICATION FROM WITHOUT

Figure 3.3
Throughout its approximately 4.5-billion-year history, the surface of the Moon has been bombarded repeatedly to form impact craters like those in this Apollo 17 view of the lunar farside. This heavily cratered surface attests to the significance of impact cratering as a planetary process.

Impact-Crater Mechanics—A Matter of Shock

As long ago as 1802, there were suggestions that some terrestrial craters might have formed by impact. The speculation that lunar craters resulted from impacts came later in the nineteenth century. It is ironic that Meteor Crater in Arizona—one of the few, well-preserved terrestrial impact craters—was not initially recognized as such. By the 1960s, geological studies had revealed about 100 probable impact craters on Earth. Nearly each year new ones still are discovered. Of course, many, many thousands of large craters have formed over geological time, but only a tiny fraction have survived the rapid erosion and burial characteristic of Earth's surface sufficiently to be still recognizable.

Impact craters result when an interplanetary object slams into a larger object, excavating a hole. The projectile's kinetic energy is transferred very rapidly to the target planetary surface, compacting its material and generating a shock wave. The shock wave, initially hemispheric in shape, expands swiftly, moving material radially. As the shock wave moves through the target, decompression occurs in its wake, generating a *rarefaction* wave, which deflects material laterally and upward.

The highest shock stresses occur within a fraction of a second of impact and transform the host material at the target into vapor, melt, and highly deformed and altered material, most of which is ejected. As the shock engulfs more and more of the target, the stresses decay radially away from the impact, and excavation takes place at lower and lower velocities until, ultimately, the strength of the target exceeds the stress of the shock, and excavation ceases.

However, the shock wave continues to expand as elastic (seismic) waves beyond the limits of the crater, imparting seismic energy far beyond the crater. Seismic jostling of terrain around impact craters may have been a significant degradational agent on planetary surfaces, at least in conjunction with large impact events.

Impacts on the Moon by bodies originating elsewhere in the Solar System probably occur at velocities typically greater than 10 kilometers per second. These are called *hypervelocity impacts*. This is also the case for the other Earthlike planets, except for very small objects, which can be slowed down significantly by the atmospheres of the Earth, Venus, or Mars.

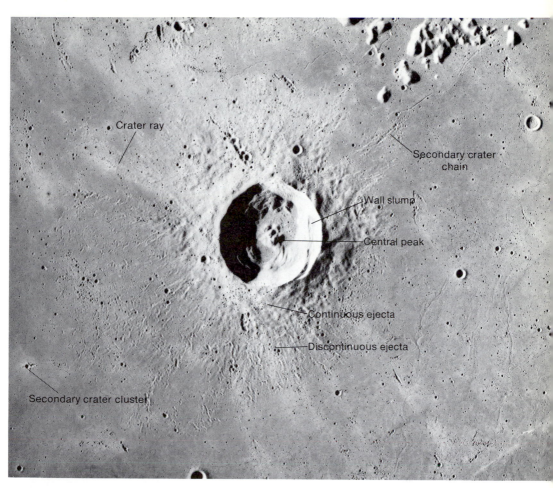

Crater ray

Secondary crater chain

Wall slump

Central peak

Continuous ejecta

Discontinuous ejecta

Secondary crater cluster

Figure 3.4
Apollo 17 view of the lunar crater Euler in southwestern Mare Imbrium. This 20-kilo-
meter-diameter impact crater shows the near-rim ejecta deposits characterized by continuous, smooth, relatively uncratered surfaces and discontinuous ejecta farther out, associated with chains of secondary craters. Bright rays can be seen extending radially from the
crater toward the upper-left hand corner of the picture.

Material thrown out of the crater—*ejecta*—is distributed radially from the crater rim, with most of the material deposited
close to the rim (Fig. 3.4). Impacts of large objects produce clots
of ejecta that themselves become impact projectiles and create
smaller impact craters called *secondary* craters. Radiating outward from the zone of continuous ejecta are *crater rays*, which

appear on photographs as high-albedo wispy filaments. Crater rays may consist of fine particles of ejecta, including impact-generated glass beads, and/or local material that has been disturbed by the formation of small secondary craters.

Regolith—Impact-Generated Soils

Throughout geological time, impacting objects of all sizes have bombarded the surfaces of the Earthlike planets. On the Moon and Mercury, and to a lesser extent on Mars (Fig. 3.5), this process has built up "soil" layers of impact-generated fragments, or *regolith*. (These soils, of course, do not include the microbiological component usually implied by the term "soil" on Earth.) The thickness of regolith on any given surface tends to increase with age. Impact craters on the Moon range in size from microcraters to great basins; the most effective size for generating regolith on the Moon appears to be craters in the range of 10 to 1500 meters in diameter, at least on mare surfaces. The reason is that larger impacts occur so infrequently as to affect directly only a small fraction of the surface, whereas smaller impacts cannot deeply penetrate the regolith formed by the intermediate-sized craters and thus simply remix the existing regolith. The very small impacts do contribute, however, to smoothing processes, particularly on planets lacking atmospheres. Closeup pictures of the lunar surface, for example, show rocks that are crumbling away as a result of hundreds of thousands of tiny impacts.

Regolith consists of rock fragments derived in place, pieces of ejecta, possibly an occasional altered fragment of a meteorite, and glass formed from impact melt. This is the "soil" of the Moon and very probably of Mercury as well. However, the regolith-generating process can also compact the soil to form a new rock, composed entirely of rock fragments and soil. Such a rock is called a breccia. Much of the material brought to Earth from the Moon by the Apollo astronauts is breccia.

Soils can be expected to be quite different on planets with atmospheres. Atmospheres consume small entering projectiles and also eliminate the sandblasting effect of secondary impacts. For this reason, impact craters smaller than a few hundred meters in diameter are unlikely to occur on Venus. Thus, instead of impact cratering, atmospheric agents, including winds and liquids, would be primarily responsible for the formation of soils, and chemical weathering is likely to produce clay minerals and otherwise dominate the soil-formation process.

Regolith observed on Mars by the Viking Landers (Fig. 3.5) includes angular blocks ranging in size from a few centimeters to more than a meter across. Most of the blocks were probably derived by impact processes and have been subsequently modified by surface agents, including winds. In addition to impact-generated fragments, the martian soil contains fine windblown sediment.

Impact Craters—Sizes, Shapes, and Features

The formation of an impact crater is an energy-transfer process. Since a projectile's kinetic energy increases with the square of its velocity, the mass of an impacting body contributes relatively less to the energy of the impact than does its velocity. Thus small objects traveling at high speeds can create surprisingly large craters. On the Moon and Mercury, objects of all sizes bombard the surface, creating a wide range of crater sizes. Moreover, lunar samples reveal minute impact craters only micrometers in diameter preserved on tiny mineral or glass fragments.

The basic shock physics of impact-cratering processes is quite similar for all crater sizes, as evidenced by data gathered by laboratory experimentation and extended to full-sized structures through theoretical scaling laws. As a consequence, the morphology of impact craters on the different Earthlike planets can be correlated with different environmental conditions. For example, the importance of gravity as a control in crater morphology became apparent in a comparison of lunar and mercurian craters. The Mariner 10 images of Mercury reveal that although its cratered surface resembles that of the Moon, there are certain dissimilarities, due, in part, to its much higher surface gravity (Mercury's gravity is 1/3 that of Earth; the Moon's 1/6 that of Earth). The band of continuous ejecta and the zone of

A

B

Figure 3.5

In these four frames, the close-up appearance of the surface of the Moon, Mars, and Venus are compared with a locale on Earth where chemical weathering has been of little consequence. **(A)** Surface of the Moon at Apollo 17 landing site showing impact-generated regolith consisting of particles of all sizes from dust to boulders. **(B)** Surface of Earth on glacial deposits in Iceland showing fine, windblown, basaltic sand (dark material) and weathered boulders (boulder in foreground is 70 cm). **(C)** Surface of Venus viewed by Venera spacecraft showing particles of predominantly two sizes: fine, subcentimeter material and cobble- to boulder-sized material (boulder in foreground is 30 cm). **(D)** Viking Lander 2 image showing block-littered surface of Mars and fine-grained windblown material (rocks in foreground are about 15 cm).

C

D

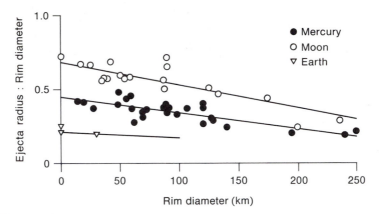

Figure 3.6
Average radial extents of continuous ejecta deposits around fresh craters on Mercury, the Moon, and Earth. Lines through mercurian and lunar data points are least-squares fits to measurements; Earth data points are based on Ries Basin, Germany; Meteor Crater, Arizona; and laboratory hypervelocity (6−7 kilometers per second) impact craters formed in noncohesive sand. [After Donald E. Gault, John E. Guest, John B. Murray, Daniel Dzurisin, and Michael C. Malin, *Journal of Geophysical Research*, vol. 80, p. 2452, 1975; copyrighted by American Geophysical Union.]

secondary craters are both much closer to the rims of the primary craters on Mercury than on the Moon (Fig. 3.6). However, if the expressions for trajectories of ejecta are corrected for the higher surface gravity of Mercury, ejecta is predicted to fall more quickly from hypervelocity impacts on Mercury and hence should be located closer to the rim, as observed. In analyzing the distribution of ejecta on Mars and Venus, it is necessary to allow for the effect of atmospheric drag, the nature of the target, and the number and size of impacting bodies consumed or slowed down by the atmosphere.

Extensive studies of crater morphology on the Moon show that it is useful to group lunar impact craters into four size classes, each with distinctive characteristics: (1) microcraters—1 centimeter and smaller in diameter; (2) small craters—1 centimeter to 15 kilometers; (3) large craters—15 to 300 kilometers; and (4) basins—craters larger than about 300 kilometers.

Rocks that show microcratering have been peppered by small particles that chip the surfaces and slowly reduce the volume;

some may also be split into smaller pieces, speeding up disintegration. These processes ultimately lead to degraded surfaces and the rounding of angular fragments.

Small craters are distinguished from larger craters by their relatively simple form. Typically, their floors and walls lack structural features; some show minor slumping. From the rim crest to the floor, the walls consist primarily of talus, plus some zones of outcrop. In profile, lunar craters smaller than 2 kilometers in diameter may be bowl-shaped, flat-floored, or stepped. Experiments suggest that the variation in profile can be attributed largely to differences in the strengths and thicknesses of layered target materials (Fig. 3.7). (The regolith layer is the top, weak, target layer in this case, and is underlain by one or more stronger units.)

Large craters have floors that typically appear to have been modified by molten rock (either impact melt, volcanic extrusion, or both). In addition, they may have a central peak, a cluster of them, or, in the largest craters, which grade into basins, a central ring or rings. Although the details of the origin of central peaks remain poorly understood, studies of the stratigraphy of terrestrial impact craters show that a plug of rock is elevated from great depth to produce central uplift, perhaps related to a "rebound" process following the excavation of the crater.

The walls of most large craters exhibit extensive modification, primarily by debris flows and slumping, which produce terraces. Igneous activity also is evident in the form of small flat "pools" of dark material occupying low regions within the wall terraces, which represent impact melt or, in some cases, volcanic activity. Outside the rim crest, large craters display complex ejecta deposits (Figs. 3.4, 3.8).

In lunar craters larger than about 120 kilometers, one or more inner rings replace the central peak. Some have as many as five rings. Lunar craters larger than about 300 kilometers in diameter are usually termed basins, although many investigators prefer to limit the application of that term to large multiringed craters.

One of the more striking results obtained by the Viking Orbiter television system was the revelation of new kinds of ejecta patterns around some martian craters. The ejecta appears to have been emplaced more as a flowing fluid mass rather than as

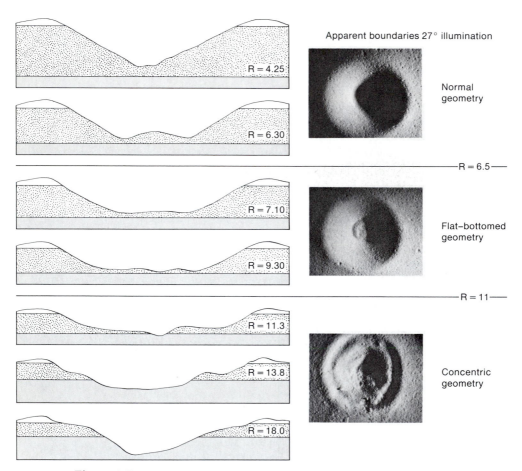

Apparent boundaries 27° illumination

R = 4.25

Normal geometry

R = 6.30

——————————————————————R = 6.5——

R = 7.10

Flat–bottomed geometry

R = 9.30

——————————————————————R = 11——

R = 11.3

Concentric geometry

R = 13.8

R = 18.0

Figure 3.7
This series of experimental craters produced at the NASA Ames Research Center Hyper-velocity Impact Facility illustrates the relation of crater morphology to thickness of "rego-lith" (loose sand) overlying a more competent target material. R is the ratio of crater diameter to regolith thickness; thus "normal" craters form in thick deposits ($R \leq 6.5$), flat-bottomed craters form when R is between 6.5 and 11, and concentric craters result from very thin regolith deposits where $R \geq 11$. The results from these experiments have been applied to estimate regolith thickness on different lunar mare surfaces. [From Verne R. Oberbeck and William L. Quaide, *Journal of Geophysical Research*, vol. 72, fig. 3, p. 4700, 1967; copyrighted by American Geophysical Union.]

Figure 3.8
Vertical view of King Crater, a 75-kilometer crater on the lunar farside showing radial flow of the ejecta deposits, central peak complex, and terraced walls. Smooth patches on the floor and on the northwestern part of the crater's outer flank (upper left) may be pools of impact melt.

ballistic debris (Fig. 3.9). A possible explanation is that the impact occurred on a surface that included ground ice. It is speculated that heat produced by the impact may have melted ice that mixed with the fragmental ejecta to produce a slurry-like mass. This mass may initially have been ejected ballistically, but, after falling to the surface, continued to travel away from the primary impact area as a surface flow.

Multiringed Basins on the Earthlike Planets

For many years, large circular lunar features such as Imbrium and Serenitatis were recognized as dominant structural elements of the crust. Nearly all the major mountain chains on the Moon are concentric about circular basins, and most of the large fracture systems in the highlands seem to be oriented around the basins. However, the difference between circular *basins* and circular *maria* was not fully appreciated until topographic and geological mapping from photographs was underway and pictures of the farside were obtained. The distribution of lunar basins appears to be the result of random, large-scale impacts. The circular maria, however, were formed later than the topographic basins they occupy. The nearside of the Moon is topographically lower than the farside (i.e., gravitationally and geometrically lower relative to the center of the mass). This asymmetry appears to be responsible for the localization of most of the maria on the nearside.

The Orientale Basin on the Moon's western limb is an especially well-exposed, well-preserved multiringed basin (Fig. 3.10). As many as four concentric mountain rings can be recognized, the most prominent being the Cordillera Mountains (930 kilometers in diameter) and the Rook Mountains (620 kilometers in diameter). The history of the ring mountains remains contentious, but their formation appears to be related to post-impact adjustments of the crust—perhaps some form of slump-block movement on a gigantic scale. Eruptions of mare basalts have partly filled the center of the basin and formed restricted pools of lava between some of the ring mountains. Examination of the terrain surrounding the Orientale Basin reveals ejecta deposits that in most respects resemble scaled-up manifestations of the ejecta of smaller craters.

Figure 3.9
These Viking Orbiter 1 pictures of the 28-kilometer-diameter crater Arandus in the Cydonia region of Mars, about halfway between the equator and the north pole, display an unusual style of cratering. The ejecta from the crater appears to have been emplaced primarily by flow of a fluidized mixture of fragmental ejecta and perhaps water derived from melted ground ice.

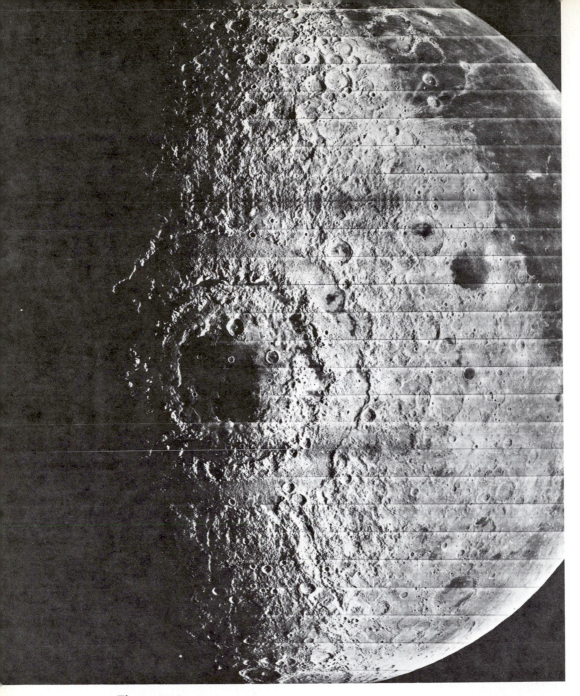

Figure 3.10
Impact craters larger than about 300 kilometers in diameter are termed basins. Shown
here is the multiringed Orientale Basin, one of the youngest and best-preserved basins on
the Moon. The outer ring of mountains, named the Cordillera Mountains, is about 900
kilometers in diameter; they contain the next two rings, the Inner and Outer Rook Mountains. The mare region to the top right is Oceanus Procellarum.

Basins are very old features on the Moon. Samples of the Fra Mauro Formation (the ejecta from the impact that formed the Imbrium Basin) have been dated at about 3.98 billion years of age. On Earth, analogous scars have been destroyed by crustal plate motion and atmospheric and hydrospheric erosion. Mars boasts a 2000-kilometer-diameter basin (Hellas) plus many smaller basins, such as Argyre and Libya. Mariner 10 revealed a 1300-kilometer-diameter multiringed crater on Mercury—since named the Caloris Basin—plus more than 20 smaller basins. Earth-based radar images of Venus suggest the presence of large circular depressions, but it is not yet known whether these are impact basins. Multiringed basins appear to be integral features of the crustal structure and early history of all the inner planets, and some of the outer planet satellites as well.

Cratered Surfaces—Keys to the Past

As a particular surface continues to accumulate impact craters, the abundance of crater scars increases until the surface becomes nearly saturated by craters of a given size and smaller (Fig. 3.11). By comparing the areal densities (surface abundance) of craters in one unsaturated region with those of other unsaturated regions on the same planet, it is possible to assign a relative age sequence to those surfaces, provided that smoothing processes have not operated differently from place to place on that planet. Carrying the idea one (very big) step further, if the average rate of impact crater formation can be estimated, it is possible to obtain absolute ages for the surfaces. Finally, to the extent that the relationships between impacting objects on one planet can be correlated with those on the other Earthlike planets, it will be possible to make interplanetary temporal correlations. For surfaces nearly saturated with craters, even of very large sizes, an early epoch is implied during which rates of crater formation were enormously greater than at present. Such conditions apparently prevailed during the early history of the Moon, Mercury, and Mars.

There are, however, many difficulties in attempting to obtain

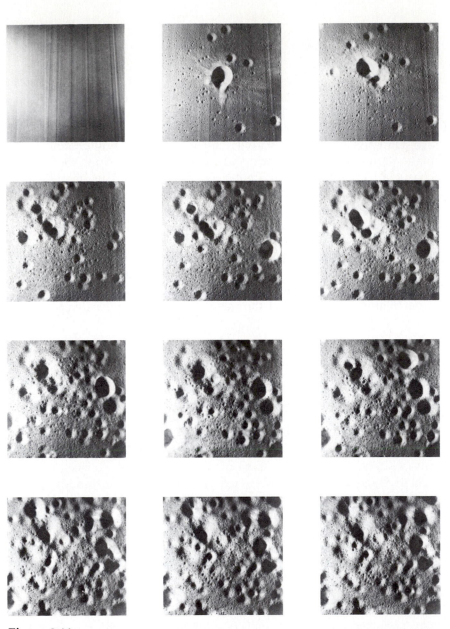

Figure 3.11
In this laboratory simulation of the development of a cratered terrain, termed *Mare Exemplum*, projectiles of six different sizes were impacted into dry quartz sand so that ten craters were formed in one size class for every one in the next-larger size class. After a period of time, the number of craters formed balanced the number destroyed, resulting in an *equilibrium* condition. Such a condition is believed to exist for small craters in the

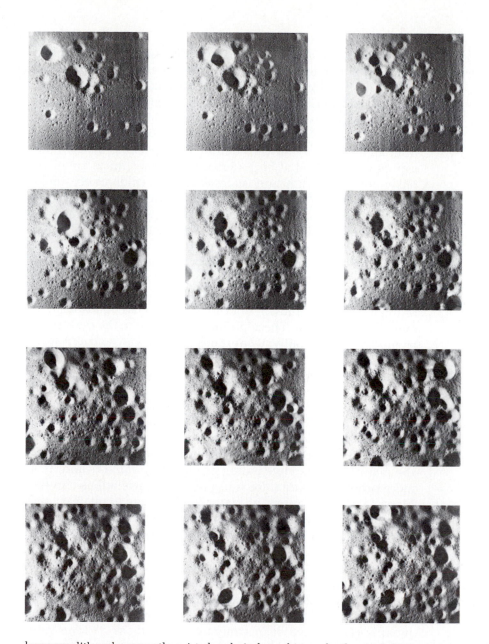

lunar regolith and apparently existed early in lunar history for the much larger craters preserved in the lunar highlands. Relative ages can be obtained by comparing crater counts on one frame with those of another frame. However, relative ages cannot be determined between two surfaces that have reached equilibrium. [After Donald Gault, *Radio Science*, vol. 5, fig. 5, p. 277, 1970; copyrighted by American Geophysical Union.]

either relative or absolute ages from crater statistics alone. The assumption is often made that all the craters examined are of impact origin. But circular depressions also can form through volcanic and subsidence processes. In fact, it is not always possible to distinguish impact craters from craters of internal origin. The morphological criteria for separating one crater type from another are not always definitive, especially if the crater cannot be examined at high resolution. Some craters are obviously not of impact origin, such as summit calderas on shield volcanoes. Except for some secondary impact craters, those craters that occur in association with volcanic features, that are irregular in planimetric form, or that occur in chains rather than randomly are probably not of impact origin.

The number and size distribution of all secondary craters associated with primary impacts must be determined in order to adjust the total crater count. For absolute dates, the frequency of impact must be known, not only for present conditions, but over geological time. These problems have led to the derivation of supplementary dating techniques that take into account various properties of crater morphology as well as simple crater counts. For example, as a crater ages, at least on the Moon or Mercury, it necessarily degrades in form through a series of stages until it is barely discernible; the distribution of morphologies thus is another indicator of relative age that can, in some instances, be exploited.

Despite the evident problems, crater statistics continue to be the mainstay for relative dating of planetary surfaces and, in a more primitive fashion, for interplanetary correlations. On the Moon, age implications of crater abundances on at least the mare surfaces have been calibrated by comparison with radiometric ages obtained from samples of those surfaces returned during the Apollo program. Extrapolations and interpolations have been made to other lunar areas. For large physiographic regions, such as the lunar highlands versus the mare plains, it appears to be possible to compare their ages with those of large physiographic regions on other planets. For example, cratered lunar surfaces have been compared with those of Mars and Mercury. However, in comparing lunar with mercurian and martian craters, adjustments must be made for differences in planetary environment and for changes in cratering rates during geological time. Moreover, the surface of Mars, especially, exhibits widely differing smoothing histories (e.g., eolian and volcanic blanketing) from place to place.

THE HANDIWORK OF GRAVITY

Gravity is important in the modification of all the planets in the inner Solar System. It produces numerous landforms on a planet's surface. Landslides and avalanches are examples of mass movements caused by gravity.

Classification and Morphology of Mass Movements

Mass movements are downward and outward movements of materials involving either rapid falling or sliding or slow subsidence or flow. On Earth, mass movements are usually classified by the rate and kind of motion, type of material, and water content (Fig. 3.12). Figure 3.13 shows the nomenclature of the parts of a landslide. Figure 3.14 shows a classification scheme that relates most types of movements known on Earth with the types of materials involved.

Mass movements on Earth are small compared to those on the other planets of the inner Solar System. Few exceed 5 to 10 kilometers in length, and almost all that do are submarine debris flows consisting primarily of sediment-laden water. Subaerial landslides on Earth are quite inconsequential compared to those on Mars. For example, Figure 3.15 is an oblique air photo, which shows a major landslide in southern California, the Blackhawk landslide. In contrast, the full Landsat image (not reproduced here), some 185 kilometers on a side, shows many morphologic features of the Mojave Desert and the San Bernardino and San Gabriel mountains; but the Blackhawk slide, one of the largest known, is nearly invisible. Even with a tenfold enlargement, diagnostic features are indistinguishable. Yet this large terrestrial slide poses a real question as to the means of lubrication involved. Presumably air trapped in the initial movement somehow helped fluidize the main landslide mass.

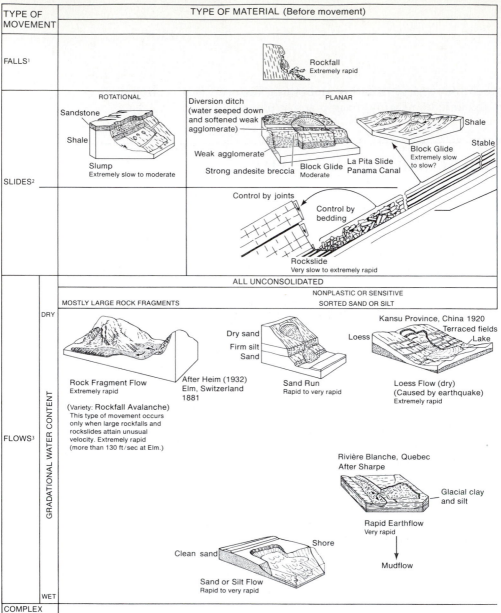

TYPE OF MOVEMENT	TYPE OF MATERIAL (Before movement)		

FALLS[1]

Rockfall
Extremely rapid

SLIDES[2]

ROTATIONAL
PLANAR

Sandstone

Shale

Slump
Extremely slow to moderate

Diversion ditch
(water seeped down
and softened weak
agglomerate)

Weak agglomerate

Strong andesite breccia

Block Glide
Moderate

La Pita Slide
Panama Canal

Block Glide
Extremely slow
to slow?

Shale

Stable

Control by joints

Control by
bedding

Rockslide
Very slow to extremely rapid

FLOWS[3]

GRADATIONAL WATER CONTENT

DRY

WET

ALL UNCONSOLIDATED

MOSTLY LARGE ROCK FRAGMENTS

NONPLASTIC OR SENSITIVE
SORTED SAND OR SILT

Rock Fragment Flow
Extremely rapid

After Heim (1932)
Elm, Switzerland
1881

(Variety: Rockfall Avalanche)
This type of movement occurs
only when large rockfalls and
rockslides attain unusual
velocity. Extremely rapid
(more than 130 ft/sec at Elm.)

Dry sand
Firm silt
Sand

Sand Run
Rapid to very rapid

Loess

Kansu Province, China 1920
Terraced fields
Lake

Loess Flow (dry)
(Caused by earthquake)
Extremely rapid

Rivière Blanche, Quebec
After Sharpe

Glacial clay
and silt

Rapid Earthflow
Very rapid

Mudflow

Clean sand

Shore

Sand or Silt Flow
Rapid to very rapid

COMPLEX LANDSLIDES[4]

[1] Mass in motion travels most of the distance through the air. Includes free fall, movement by leaps and bounds, and rolling of rock and debris fragments without much interaction of one fragment with another.

[2] Movement caused by finite shear failure along one or several surfaces which are visible or whose presence may reasonably be inferred.
A. Material in motion not greatly deformed.
Moving mass consists of one or a few units. Maximum dimension of units is greater than displacement between units. Movement may be controlled by surfaces of weakness such as faults, bedding planes or joints.
(1) *Slump:* Movement only along internal slip surfaces, which are usually concave upward. Backward tilting of units is common.
(2) *Block glide:* Movement of a single unit out and down along a more or less planar surface of weakness, generally a bedding plane. Block may glide far out on original ground surface.
B. Material in motion is greatly deformed or consists of many semi-independent units.
Movement frequently is structurally controlled by surfaces of weakness such as faults, joints, bedding planes, variations in shear strength between layers of bedded deposits, or by the contact between firm bedrock and overlying detritus. Maximum dimension of units is comparable to or less than displacement between units, and generally much smaller than displacement of center of gravity of the whole mass. Movement may progress beyond original slip surface so that parts of mass slide over the ground surface.

[3] Movement within displaced mass such that the form taken by moving material or the apparent distribution of velocities and displacements resemble those of viscous fluids. Slip surfaces within moving material are usually not visible or are short-lived. Boundary between moving mass and material in place may be sharp or a zone of distributed shear.

[4] Movement is by a combination of one or more of the three principal types of movement described above. Many landslides are complex (e.g., failures by lateral spreading and rock fragment flow), although one type of movement generally dominates over the others at certain areas within a slide or at a particular time in the evolution of a slide.

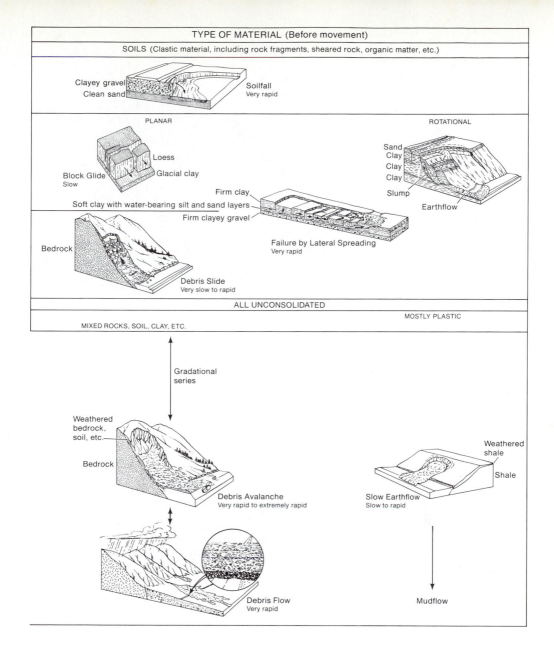

Figure 3.12
Classification of mass movements on the basis of rate and kind of motion, type of material, and water content. [Courtesy of National Academy of Sciences, National Research Council. Publication 544, 1958.]

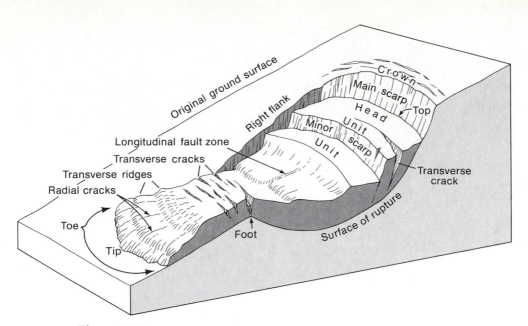

Figure 3.13
Nomenclature of the parts of a landslide. **Main scarp.** A steep surface on the undisturbed ground around the periphery of the slide, caused by movement of the slide material away from the undisturbed ground. The projection of the scarp surface under the disturbed material becomes the surface of rupture. **Minor scarp.** A steep surface on the disturbed material produced by differential movements within the sliding mass. **Head.** The upper parts of the slide material along the contact between the disturbed material and the main scarp. **Top.** The highest point of contact between the disturbed material and the main scarp. **Foot.** The line of intersection (sometimes buried) between the lower part of the surface of rupture and the original ground surface. **Toe.** The margin of disturbed material most distant from the main scarp. **Tip.** The point on the toe most distant from the top of the slide. **Flank.** The side of the landslide. **Crown.** The material that is still in place, practically undisturbed, and adjacent to the highest parts of the main scarp. **Original ground surface.** The slope that existed before the movement took place. **Left and right.** Though compass directions are preferable in describing a slide, if "left" and "right" are used they refer to the slide as viewed from the crown. [Courtesy of National Academy of Sciences, National Research Council. Publication 544, 1958.]

Figure 3.15
Oblique aerial photograph of the Blackhawk Landslide on the north flank of the San Bernardino Mountains (background) in southern California. The average size of the fragments making up the mass is less than 3 centimeters; fragments consist mostly of altered gray crystalline limestone. The thickness of the mass near the outer margin (foreground) is probably at least 30 meters. [From *Geology Illustrated* by J. S. Shelton. W. H. Freeman and Company. Copyright © 1966.]

TYPE OF MOVEMENT		TYPE OF MATERIAL			
		BEDROCK		SOILS	
FALLS		Rockfall		Soilfall	
SLIDES	FEW UNITS	ROTATIONAL	PLANAR	PLANAR	ROTATIONAL
		Slump	Block glide	Block glide	Block slump
	MANY UNITS		Rockslide	Debris slide	Failure by lateral spreading
FLOWS		ALL UNCONSOLIDATED			
		ROCK FRAGMENTS	SAND OR SILT	MIXED	MOSTLY PLASTIC
	DRY	Rock fragment flow	Sand run	Loess flow	
				Rapid earthflow	Debris avalanche	Slow earthflow
	WET		Sand or silt flow	Debris flow	Mudflow
COMPLEX LANDSLIDES		Combinations of materials or type of movement			

Figure 3.14
Classification of landslides. [After C. F. Stewart Sharpe, *Landslides and Related Phenomena*, Cooper Square, Inc., New York, 1968.]

Mass Movement on the Earthlike Planets

The Moon exhibits various features of mass movements on a scale generally similar to those on Earth. The most well developed and best known are the terraced inner walls of large craters, which display many clear-cut indicators of slumping (Fig. 3.16). These terraces may have resulted from removal of basal support accompanying crater excavation, from an increase in the slope and weight of the accumulated crater ejecta, or from the seismicity of the impact. Many lunar areas show tracks and scars left where large boulders have slid, rolled, and bounded down crater walls and other slopes (Fig. 3.17). Some craters exhibit slides of flows of loose regolith. Features developed on abnormally precipitous slopes look much like terrestrial rockfall avalanches.

In contrast, Mars exhibits numerous forms of mass movements that are of enormous scale compared to terrestrial or lunar landforms. Massive slides, slumps, and flows are clearly visible in moderately low-resolution spacecraft images; diagnostic features are visible in some high-resolution Viking Orbiter photographs (Fig. 3.18).

A characteristic mass-movement landform on Mars consists of parallel, U-shaped chutes or troughs extending directly downslope from precipitous brinks, neither branching nor joining (Fig. 3.18). The configuration of crown and main scarp, the absence of head and toe, and the termination in a smooth, gentle slope resembling a skree all suggest rockflow or debris avalanche. Other even larger features have longitudinal lineations,

Figure 3.16
Oblique view of the 18-kilometer lunar crater Dawes obtained from orbit by the Apollo astronauts, showing extensive talus and other mass-wasted material on the floor of the crater. Immediately below the crater rim is an outcrop of apparently consolidated bedrock.

Figure 3.17
Boulders and boulder tracks on the North Massif photographed by the Apollo 17 astronauts with a telephoto lens from the landing module. The prominent boulder on the right is many meters across and left a track that can be traced nearly 1000 meters uphill (out of view in this picture). Illumination is from the right.

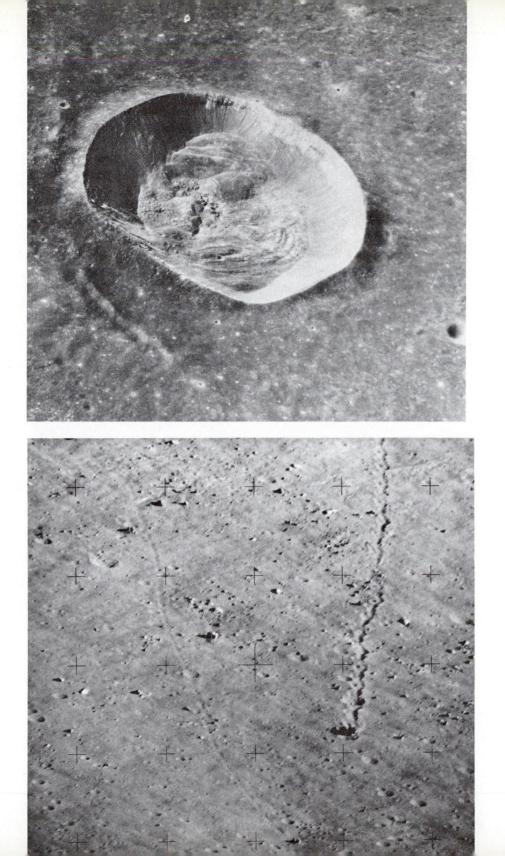

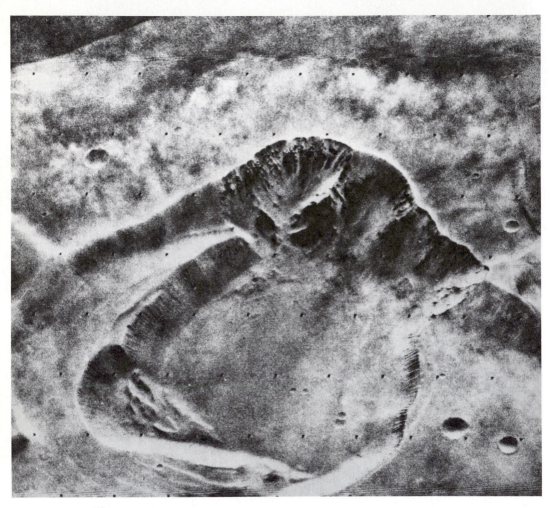

Figure 3.18
This slightly oblique view of the summit caldera of Tharsis Tholus, at 13.4°N latitude and 90.7°W longitude, shows parallel avalanche chutes on the right wall and slump cracks and slump blocks. The caldera is nearly 50 kilometers long.

Figure 3.19
This view was taken by Viking Orbiter 1 on July 3, 1976 from a range of 2000 kilometers (1240 miles) looking southward across Valles Marineris—Mariner Valley. The area shown is 70 kilometers (43 miles) by 150 kilometers (94 miles). Aprons of debris on the canyon floor indicate how the canyon has enlarged itself. The walls appear to have collapsed at intervals to form huge landslides that flowed down and across the canyon floor. Linear striations on the landslide surface show the flow's direction. On the canyon's far wall, one landslide appears to have ridden over a previous one. Streaks on the canyon floor, aligned parallel to the length of the canyon, are evidence of wind action. Layers in the canyon wall indicate that the walls are made up of alternate layers of rock, possibly lava and ash or windblown deposits.

transverse ridges, flow lobes, and irregular toe morphology suggestive of slump and flow (Fig. 3.19). Along a major scarp that outlines the circumference of the huge volcanic mountain Olympus Mons, one clearly defined region shows remarkable similarity to the Blackhawk slide, although the horizontal scale is larger by a factor of about five. Slump blocks are seen as terraces not only within impact craters but also within volcanic calderas, along the margins of chaotic areas believed to be formed by collapse, along the Olympus Mons escarpment, and along the walls of the major troughs that cross the equatorial regions.

Viking observations give new insights into mass movements on Mars. High-resolution images of large landslides show characteristics nearly identical to terrestrial analogs. Longitudinal troughs and ridges, crisp crown and main scarp, and backward rotation of slump blocks are clearly seen. Layers are preserved in both the major and minor scarps (Fig. 3.19).

On Mercury, where there is very little photographic coverage of the surface at a resolution better than even half a kilometer, the only large-scale mass movements recognized so far are the slump terraces inside large craters and some possible slumps or slides along crater walls or down the large mercurian escarpments (Fig. 3.20). Mass-movements features on a martian scale seemingly are absent.

Questions Raised by Mass Movement on the Earthlike Planets

Two major questions arise concerning mass movements on the Earthlike plants. First, how can these movements resemble terrestrial landslides in length without the existence of a lubricating airlike fluid? Second, what mechanisms initiate the movements?

Comparisons between the Earthlike planets can provide insight into the mechanisms and materials active there. For example, many features seen on both the Moon and Mars clearly indicate that weakly coherent or unconsolidated materials participate in the mass movement. However, the great distance of material transport on Mars remains enigmatic. On the basis of field observations, it has been suggested that the Blackhawk and similar large terrestrial avalanches owe their great length

Figure 3.20
Mass movement on Mercury is illustrated (arrow) in this high-resolution Mariner 10 photograph showing conformal blocks resembling glide blocks of crater wall material, with movement probably precipitated by the impact of the smaller crater on left side. Illumination is from the left.

to a lubricating layer of air that becomes trapped and compressed beneath the falling rock mass as its vertical motion is transformed to horizontal motion. Just what hypothetical martian conditions could plausibly explain the much larger martian features analogously is an unmet scientific challenge.

Lunar slides are remarkably analogous to these terrestrial features, as shown in Figure 3.21. This plot shows the relationship between the theoretical potential energy released by each slide's movement and the ratio of slide length to initial height. It is thus a graphical representation of slide efficiency. Some

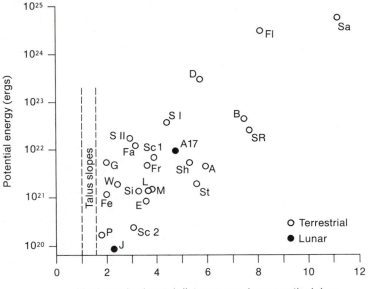

Figure 3.21
Efficiency of landslides is evident if the potential energy of large rock avalanches is plotted against the ratio of the maximum total horizontal distance traveled to the maximum total height of fall. The data are approximate but clearly show a trend toward increased efficiency with increased energy. Terrestrial avalanches are identified as follows: **A**, Allen; **B**, Blackhawk; **D**, D'Ousoi; **E**, Elm; **Fa**, Fairweather; **Fe**, Fernandina; **Fl**, Flims; **Fr**, Frank; **G**, Goldau; **L**, Little Tahoma Peak 3; **M**, Madison; **P**, Puget Peak; **Sa**, Saidmerreh; **S I**, Sawtooth Ridge I; **S II**, Sawtooth Ridge II; **Sc**, Schwan 1; **Sc 2**, Schwan 2; **Sh**, Sherman; **SR**, Silver Reef; **Si**, Sioux; **St**, Steller; **W**, Wolf. Lunar avalanches are: **A17**, Apollo 17; **J**, Jansen B crater. [After K. Howard, "Avalanche Mode of Motion: Implications from Lunar Examples," *Science*, vol. 180, pp. 1052–1055, fig. 3, June 8, 1973. Copyright 1973 by the American Association for the Advancement of Science.]

form of dry fluidization has been proposed to account for the high efficiency. Martian landslides are similarly involved in this controversy, since their efficiencies are comparable to both lunar and terrestrial examples. If a cushion of gas is required, it implies that volatiles were present not only on Mars but on the Moon as well, where other evidence strongly argues against their presence. Thus there appears to be a dilemma concerning large landslides on these planets.

Some downslope movement is precipitated by material thrown out of craters. It is possible that these kinetically initiated landslides owe part of their length to this ballistically derived material. The paucity of post-mare landslides suggests that many mass movements probably occurred early in lunar history. In contrast, slides and slumps on Mars are associated with fresh, youthful features (e.g., volcanic calderas, canyon walls), suggesting the possibility of significant past seismic activity. Since wind can both undermine slopes and deposit new material above them, it too must be considered as an initiator of landslides.

There is much more to be learned about the role of gravity in altering planetary surfaces. We already know that it has a marked effect, usually working in collaboration with other processes.

WIND—THE PERSISTENT SCULPTOR

Sandstorms and other eolian processes have long been recognized as important in shaping the Earth's surface. The significance of eolian processes on other planets was not fully appreciated until the Mariner 9 mission to Mars (1971–1972) revealed abundant eolian features. Prior to Mariner 9 some observers believed that the very high winds required by the thin atmosphere to move particles on Mars did not occur frequently enough, if at all, to be geologically important.

Similarly, eolian processes were thought to play a limited role on Venus until the Soviet Venera 9 and 10 spacecraft returned pictures of its surface in December 1975. Circulation models based on observations of the upper atmosphere, when extrapolated to the surface, indicated that near-surface wind velocities should be very low because of Venus' dense, viscous atmosphere. However, wind measurements made at the surface by Venera 9 and 10 suggest that, under some circumstances, winds there are capable of moving sand-sized particles.

The Moon and Mercury lack atmospheres and, therefore, winds, nor is there any evidence to suggest that circumstances were different in the past. Thus only on the Earth and Mars are eolian processes definitely of major importance. Data from the 1979 Pioneer Venus probes indicate a reasonably clear atmosphere near the surface, suggesting that dust transport on Venus was not a significant process at that time.

The Movement of Particles by Wind on Earth, Mars, and Venus

Particles are transported by wind in three ways: traction, saltation (jumping), and suspension (Fig. 3.22). Both traction and suspension are generated by the impact of saltating particles.

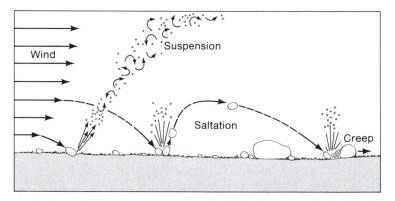

Figure 3.22
The friction of the wind movement across a surface can set particles into motion if the wind is sufficiently strong. The diagram illustrates how particles are moved by the wind in three modes: suspension, saltation, and creep. [From Thomas A. Mutch, *The Geology of Mars* (copyright © 1976 by Princeton University Press): fig. 1, p. 236. Reprinted by permission of Princeton University Press.]

Intuitively, the higher the wind velocity, the more likely that particles will move. It is also intuitive to expect smaller particles to require lower wind velocities to move them than larger particles. However, field observations and laboratory tests reveal a counterintuitive circumstance. On Earth, for particle diameters smaller than about 80 microns (sand sized), *stronger* winds are needed to initiate new motion. The reason for this is not fully understood, but it is partly the result of aerodynamic effects and the increase in cohesion that accompanies decrease in particle size. The relation of grain size to threshold is shown in Figure 3.23, where it can be seen that for any fluid density there is a particular grain size that can be moved by the least wind.

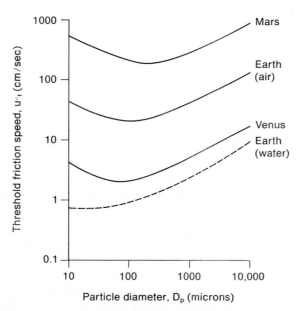

Figure 3.23

The threshold friction wind speed (minimum wind to set particles into motion) is shown as a function of particle diameter for Mars, Earth, and Venus. The lowermost curve shows the minimum water velocity required to set particles into motion. The very low atmospheric density on Mars requires much higher winds to move particles than on Earth. The higher atmospheric density on Venus results in much lower wind speeds. [After J. D. Iverson, Ronald Greeley, and James Pollack, *Journal of the Atmospheric Sciences*, vol. 33, fig. 6, p. 2428, 1976; copyrighted by the American Meteorological Society.]

Laboratory experiments done under simulated martian atmospheric pressures (0.1 to 1.0 percent that on Earth) show that the wind velocities required to move particles on Mars range from about 90 to 240 kilometers per hour, depending upon the atmospheric pressure at the surface in question and on surface roughness. The particle size most easily moved by the wind on Mars is about 160 microns in diameter. On Venus, by contrast (pressure about 100 times that on Earth), much lower wind velocities are needed, and the particle size most easily moved is smaller than that moved by the least wind on Earth.

The exact mechanism for initiating particle movement by the wind is poorly understood. However, once particles begin to saltate, they strike other particles. Particles too small to be moved by wind alone are struck by saltating grains and, once moved, may be transported in suspension or even carried aloft into the free stream; thus do dust storms arise. Particles too large to be moved by the wind alone, such as fine gravel, may also be struck by saltating grains and thus be pushed along the surface by *saltation creep*.

In comparing the effects of eolian erosion on these different planets, we must remember that many factors are involved. Even though the velocities needed to move particles on Venus are relatively low, those velocities may not occur very frequently. Most important, its dense atmosphere also will have a cushioning effect that tends to decrease the effectiveness of erosion and the generation of windborne particles by saltation. In contrast, this effect is minimal in the thin martian atmosphere. Once particles begin to move on the martian surface, the higher wind velocities impart more energy to the particles, giving them a higher capacity for abrasion than on the Earth or Venus. Thus, in comparison with the Earth, where agents of modification associated with water tend to overshadow eolian processes, and in comparison with the surface of Venus, where the atmospheric-surface interface must be more like that of an ocean bottom, Mars stands out as the planetary surface of spectacular eolian features.

Dust Storms and Eolian Features on Mars

Earth-based telescopic observations of Mars over the past 150 years have revealed yellow "clouds" that move over the planet's disk. The clouds seem to originate consistently in particular geographic localities and occur most frequently during perihelion, when Mars receives maximum solar heating. Long ago, such observations led to speculation that the yellow clouds were actually giant dust storms. This idea was confirmed by the instruments aboard the orbiting spacecraft Mariner 9 in 1971–1972, when a major dust storm was observed to completely envelope the planet, obscuring the surface from the spacecraft's cameras.

Dust storms typically begin in Hellespontus, Noachis, and Solis Lacus, three elevated plateaus in the southern hemisphere between 20° and 40°S latitude. A storm is first manifested (to an Earth-based telescopic observer) as bright "spots" or "cones" less than 400 kilometers in diameter that last about five days. In the next 35 to 70 days, the storm expands into secondary "cones" that grow until the planet is entirely obscured. Finally, the storm slowly subsides, the polar regions being the first areas to clear. This last phase is 50 to 100 days long. Mariner 9 observations showed that the dust in the cloud was well mixed up to an altitude of 30 to 40 kilometers. Particles averaged about 2 microns in diameter, about the same size as particles in major dust storms on Earth.

As the 1971 martian dust storm settled, more and more surface became visible from Mariner 9, revealing many features attributable to eolian processes. Even after the major storm subsided, several small local wind storms took place, causing change that could be seen on the surface. Yet observations on the surface during the 1977 dust storm by the Viking Landers showed very little effect, even though simultaneous Viking Orbiter photographs revealed heavy surface obscuration. Thus the long-term erosional and depositional role of the global dust storms remains uncertain from these observations.

The most abundant eolian features on Mars are bright and dark streaks associated with craters and hillocks (Fig. 3.24). The outlines of dark streaks have been observed to change within a few weeks, whereas the bright streaks generally remain constant for months or more. Experimental comparisons (Fig. 3.25) with the martian streaks suggest that most dark streaks are the result of scouring away of particles (probably deposited during a preceding major dust storm). At least some bright streaks probably are deposits of dust built up by wind interacting with prominent craters. Some dark areas on the floors of large craters and in other topographic depressions proved to be sand dune fields (Fig. 3.26). These dune fields are comparable in size and form to large ones on Earth and are evidence of saltating sand transport and deposition on Mars.

Many martian terrains appear to be modified by eolian processes, both erosional and depositional. A region in the mid-northern latitudes appears to have been eroded, exposing a polygonally fractured surface. The ejecta of craters in this region has been partly eroded. In other regions, cratered plains are characterized by a high frequency of shallow, flat-floored craters. These regions appear to be mantled, perhaps by eolian deposits.

Many other eolian features were observed by the Mariner 9 and Viking orbiters, including the layered terrain of the polar region, "blow-outs" and other deflation features, and kilometer-scale fluting and grooving.

Figure 3.24
Light and dark albedo patterns can be associated with martian craters, as shown in these frames. The albedo patterns are considered to be the result of eolian activity. Dark streaks (left-hand side of figure) are probably eroded zones where the surface has been swept free of windblown particles; light streaks may be deposits of windblown particles. [From "Crater Streaks in the Chryse Planitia Region of Mars: Early Viking Results," by R. Greeley, R. Papson, and J. Veverka, *Icarus*, vol. 34, fig. 1, 1978. Academic Press, Inc.]

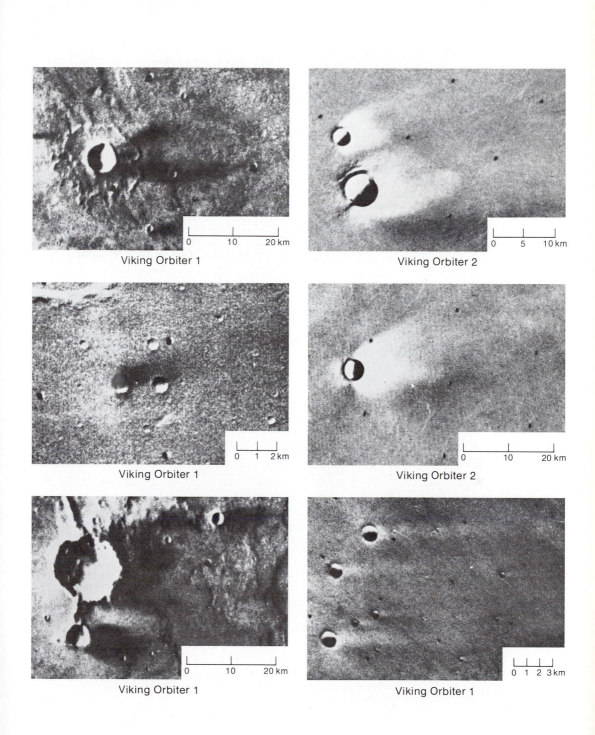

Viking Orbiter 1

Viking Orbiter 2

Viking Orbiter 1

Viking Orbiter 2

Viking Orbiter 1

Viking Orbiter 1

Figure 3.25 *(facing page)*

(A to C) Sequential photographs of a 17.8-centimeter crater modeled in loose sand and placed in a wind tunnel with a wind velocity of about 420 centimeters per second. (Illumination is from upper left and wind from left to right, as indicated by the arrow.) The crater became ovoid in outline, pointing upwind, and developed two erosional depressions off the sides of the rim that resulted from a vortex being shed from the rim. **(D)** Small martian crater in Mare Tyrrhenum showing similar outline and dark zones off the leeward edge of the crater rim that are interpreted to be the result of erosion. **(E to G)** Sequential photographs of a 17.8-centimeter crater modeled in solid wood (nondeformable), partly buried by loose sand **(E)** and subjected to a wind of 850 centimeters per second until relatively stable conditions ensued **(G)**, in which the model surface was swept free of loose sand except in zones of relative desposition (shown by the white trilobate pattern and the white patch on the windward rim). **(H)** A 2-kilometer martian crater in the region northwest of Memnonia (showing a similar trilobate pattern in the immediate lee of the crater and a white zone on the windward rim that are interpreted to be eolian deposits) and a large dark zone in the crater wake interpreted to have resulted from erosion. [After R. Greeley, J. Iversen, J. Pollack, N. Udovich, and B. White, *Science*, vol. 183, fig. 2, p. 848, 1974. Copyright 1974 by the American Association for the Advancement of Science.]

Figure 3.26

A sand dune field on the floor of a 60-kilometer crater near the south polar region of Mars (about 68°S, 215°W) is shown in this Viking Orbiter frame. The dune field is one of many such fields in this region and exhibits many similarities to terrestrial dune fields. Illumination is from the left.

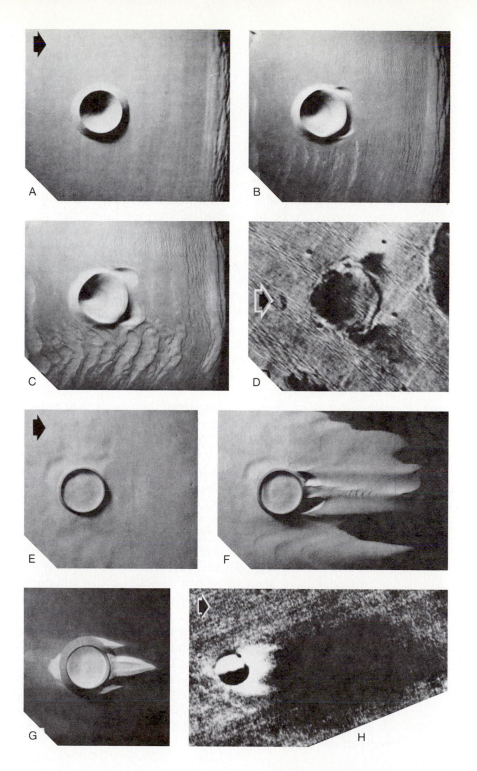

Mars' Polar Regions

The polar regions of Mars display a large variety of eolian features and the consequences of cyclical variation in eolian processes. At least three similar stratigraphic units, all probably of eolian origin, can be recognized in both the north and south polar regions. From youngest to oldest, these are: (1) a thin mantling deposit that begins at about 30° to 40° north and south latitudes and extends poleward, (2) thin, uniformly layered deposits that erode to produce characteristically terraced terrain, and (3) a massive deposit in which pitted and etched terrain develops. Also present are older units not of eolian origin, including an ancient, heavily cratered surface in the south polar region. Capping all the units are the seasonal accumulations of frozen carbon dioxide and the permanent ice caps composed of water ice and possibly also of some form of solid carbon dioxide.

Layered terrain (Fig. 3.27) consists of thin, uniformly bedded layers that have been wind-sculptured to produce smoothly rounded and terraced scarps and hills. The individual beds appear to be 10 to 50 meters thick and have been speculated to consist of a mixture of ice particles and windblown dust deposited cyclically. The total volume of these thinly layered deposits in both the north and south polar regions is estimated to be about 5×10^6 cubic kilometers, or enough to cover the entire surface of Mars to a depth of 35 meters. (Although this may seem to be a vast quantity, it is only about one-fifth the volume of the Antarctic ice cap.) Several potential sources for the fine dust material have been proposed, including the equatorial canyonlands and fretted terrain, both of which show evidence of wind erosion.

The thinly layered deposits overlie an older unit that consists of one or more massive beds in which erosion leads to the formation of pitted and etched terrain. These beds may also be of eolian origin and, like the thinly layered deposits, are also subject to eolian erosion. In some places the pitting and etching extends all the way through the unit, which averages 500 meters in thickness, to expose an underlying, older, heavily cratered surface.

Figure 3.27
Polar layered terrain (top half of picture) evidently unconformably overlies pitted terrain (lower half of picture). Layered terrain is the erosional expression of thinly-bedded eolian deposits. Each bed is perhaps 10 to 50 meters thick and may be composed of ice particles and dust.

The pits range from one-half to several tens of kilometers across and are up to 500 meters deep. Where the pits occur in groups, the terrain is described as "etched." Etched terrain is more extensive toward the equator, where the massive unit appears to be thinnest.

The material being eroded by the wind from both the thinly layered deposits and the massive deposits appears to be transported from the polar regions and redeposited as a mantling blanket toward the equator, to about 30° or 40° latitude. This mantling deposit is especially evident in small craters 1 to 10 kilometers in diameter that occur on older cratered terrains and plains. The small craters have sharp rims, and their floors appear to have been built up by infilling. In some, the floor is nearly level with the top of the rim.

Thin mantling deposits are found covering both old and young terrains. Thus this unit must be the result of the most recent geologic (climatic) variations on Mars. It has been speculated that the equatorial mantle-free zone migrates moderately northward and then southward again on a cycle of 50,000 years, in response to alternations in summer-winter solar heating of this period are believed to arise from periodic variations in Mars' orbit and spin.

Previous Eolian Regimes

Several lines of evidence lead to the conclusion that the past climate on Mars probably differed at times from the present one. For example, what caused the cyclic deposition that produced the thin, layered sediments? And what subsequent change caused these extensive sedimentary deposits in the polar regions to be now eroded and transported away from the poles? A key—and unique—aspect of Mars is the solid/gaseous exchange of the dominant constituent of the atmosphere, carbon dioxide. Changes in average polar temperatures have a profound effect on the amount of carbon dioxide in the atmosphere. Even the yearly alternations of summer and winter transfer from one-fourth to one-sixth the total volume of atmospheric carbon dioxide gas to be released to the atmosphere. As is evident from Figure 3.23, such an increase must greatly enhance eolian processes.

THE ENIGMA OF THE MARTIAN CHANNELS

Nearly every part of the Earth's surface shows the effects of running water—from the Grand Canyon to small glacial streams above the Arctic Circle. Until 1971 it was believed that this form of erosion was limited to the Earth. Then Mariner 9's view of channels on Mars radically altered this belief; subsequently, Viking orbital photography provided much high-resolution documentation of fluvial-like features on that planet.

Of principal interest are the long, narrow, linear or sinuous depressions found predominantly in the martian equatorial regions. Figure 3.28 shows part of the regional concentration of large channels on Mars. The vast majority of erosional channels lie within the heavily cratered area, and many are associated with margins of the cratered terrains.

Categories of Martian Channels

Each martian channel displays unique characteristics. Channels can be categorized according to length, width, and depth, floor and wall characteristics, head and distal terminations, relationships to other channels, relative age and topographic, geologic, and geographic setting. Alternatively, channels or channel-like forms can be categorized (genetically) as the inferred products of specific processes, although not confidently from orbital photography alone.

We recognize three types of large martian channels formed by erosion: *outflow, runoff,* and *fretted.* Additionally, a fourth class of smaller channels, resembling gullies or furrows, are termed *small valley networks,* and can also be attributed to erosion.

Outflow channels are mostly large features that apparently started from localized sources. They are often broadest and deepest at the head and decrease in size distally. A decrease in relief is often associated with merging of the outflow channel onto a plains region. Some appear to be scoured, and they display features suggestive of massive catastrophic flooding. An

Figure 3.28
This oblique view northward toward the Chryse Planitia region of Mars was obtained by
Viking Orbiter 1 from an altitude of more than 30,000 kilometers. Several large channels
can be seen emptying into Chryse Planitia. Ganges Chasma, part of the Valles Marineris
complex, is in the lower right. The channels displayed here are of the outflow type that
originate in "chaotic" terrain.

example of the outflow category is Mangala Vallis (Figs.
3.29, 3.30).

Runoff channels typically start small and increase in size and
depth distally. The headwaters usually have tributaries, although these are often short and deep and not at all reminiscent
of terrestrial river tributaries. Two prime examples are Ma'adim
Vallis (Fig. 3.31) and Nigral Vallis.

MODIFICATION FROM WITHOUT

Figure 3.29
The Mangala Vallis outflow channel extends northward from an area of subdued, hummocky topography and terminates at the southern margin of a large, mare-like plain. The length of the channel exceeds 200 kilometers. The width varies from 5 to 20 kilometers (averaging 10 kilometers), and the depth is estimated to be from 500 to 1000 meters. In the northern third of the main channel, scoured material may have been deposited and partly eroded. A gentle northerly slope, decreasing to the north, is consistent with the arrangement and morphology of channel features, suggesting a northward flow of the catastrophic floods that evidently produced this feature billions of years ago.

Figure 3.30
A portion of Mangala channel imaged at high resolution shows terracing of some small
"islands" and of the resistant knobs that top these "islands."

Figure 3.31
A martian runoff channel is shown in this Mariner 9 mosaic of Ma'adim Vallis in the southern cratered terrain. Tributaries are visible in the headward region (bottom of image). "Runoff" channels typically enlarge in width and depth downstream. The channel shown here is about 600 kilometers long.

Fretted channels have steep walls with wide, smooth, concordant floors. The configuration can be complex, with irregularly indented walls and integrated craters. Control of their development by linear crustal structures, presumably fractures, is often evident. Isolated butte- or mesa-like outliers are common, and integration of adjacent channels has occurred locally. The best examples are found in the extensive area of fretted terrain in the Deuteronilus-Protonilus region. Like fretted terrain itself, these channels have wide, smooth floors and abrupt, steep walls. A large unnamed channel opening out into Ismenius Lacus is one of the best examples (Fig. 3.32).

Figure 3.32
A martian fretted channel is shown in this Viking Orbiter view of the Ismenius Lacus region. The fretted channel exhibits steep walls, flat floors, and angular channel segments. The picture covers an area about 190 kilometers by 190 kilometers.

Small valley networks are mostly 3 to 5 kilometers wide and hundreds of kilometers long. They are characterized by their network configurations which resemble, but do not mimic, terrestrial valley systems. The diversity in these landforms is striking. Some are long, narrow, sinuous, and filamental; others are short, wide, broad-floored, and stubby. Radial patterns are sometimes found around large impact basins. Amphitheater heads on both main and tributary valleys are abundant (Figs. 7.8, 7.9).

Age of the Martian Channels

The age of the martian channels can be established by studying their relationships to other landforms on Mars. One approach utilizes the relative abundance of impact craters. On this basis, it was first believed that the channels are extremely young, because of their fresh appearance and the paucity of superposed impact craters. Preliminary crater counts seemed to indicate ages similar to those of the Tharsis volcanoes and the south polar layered deposits, suggesting that Mars had undergone recent geological and, presumably, climatological changes. Initial estimates of absolute age, based upon a preliminary estimate of impact flux (which was presumed greater than that on the Moon by perhaps a factor of 25), ranged from tens of millions to hundreds of millions of years. However, current interpretations favor a lower impact flux, and include previously unrecognized resurfacing of channel bottoms. As a result, current estimates of channel ages, based on crater counts, suggest that the martian channels are older than the most heavily cratered (and thus oldest) Tharsis lava plains, that they are about the same age as the Lunae Planum lava plains, but that they are younger than the intercrater plains of the heavily cratered terrain.

An independent way to estimate the relative ages of the channels is to observe the superposition of plains materials. For example, Mangala Vallis appears to be buried in places by rather heavily cratered plains and, hence, must be older than those plains. The observation that large channels are limited to areas of heavily cratered terrain argues that they are ancient features.

It seems likely that many martian channels are older than some of the oldest volcanic plains, and thus could be among the older topographic features still recognizable on Mars, other than the heavily cratered terrains into which they are incised.

Formation of the Martian Channels

The most popular mechanism proposed for creation of the martian channels—and the one first suggested—is that of erosion by a flowing liquid, presumably water. Alternatives such as erosion by extremely fluid lavas have been investigated, but none can be as widely applied as that of flowing water. The possibility that the channels result from crustal extension has also been explored. Since numerous other martian features do indeed strongly resemble landforms developed by tension on both the Earth and the Moon, it seems rather *ad hoc* to apply this mechanism to the channels, which do not resemble terrestrial tensional features. It appears, therefore, that flowing water offers the best potential for explaining many of the martian channel features.

However, two principal factors still present great problems to the flowing-water mechanism. First, the present martian environment is much too harsh to permit flowing water; water is at present only an extremely minor constituent in Mars' atmosphere, and is unstable on the surface in the liquid phase. Even very salty water, with freezing point depressed, would be frozen most of the time on Mars. Second, the size of the martian channels indicates that tremendous volumes of water would have been required to produce them.

These problems have motivated numerous investigators to suggest cyclical climatic changes that somehow release subterranean water supplies. It has even been suggested that water stored within the ground as a mixture of solid carbon dioxide and ice (clathrate) may have been unlocked by sudden release of overburden pressure. Water adsorbed onto fine particles in the soil has been considered as another possible source. So also has water released from ground storage (e.g., ground ice) by impact or by magmatic intrusion. Several authors even have speculated on the possibility of rainfall to explain the dendritic pattern of several channel systems. That, of course, would require a drastic change in global atmospheric conditions, as well as much more water than is now available.

Intermittent cataclysmic flooding billions of years ago is perhaps the *least unsatisfactory* explanation at present. The preponderance of irregular, hummocky, and chaotic terrain located headward of the channels suggests ground seepage as a source for the water. If this water ponded, occasional catastrophic floods could have been released, producing outflow channels. In places where ponding did not occur, subsurface water may have become integrated with surface drainage, undercutting cliff walls, and producing runoff channels. The great age of the channels suggests that the channeling process is not currently active, thus implying that current environmental restrictions may not apply. Perhaps most important, many channels have features suggestive of eolian erosion, slumping, landsliding, and other processes of landscape deterioration that are collectively termed fretting. Such multiple origins have been proposed for some channels.

Calculations based on the most erosive terrestrial floods suggest that enormous amounts of water would have had to flow down the channels seen on Mars if they are of fluvial origin. Hypothetically, water stored in the polar deposits in the form of ice (and that hypothetically acting as the cement of the layered sedimentary units), can account for perhaps only a fraction of the amount required. Unless enormous amounts of water have been lost from the planet, or somehow reburied beneath the surface, recycling of water must be invoked on a vast scale; the only alternative is that other mechanisms must have formed channels that were subsequently modified. It is this kind of dilemma—where *no* completely satisfactory explanation is apparent—that suggests that the full explanation of the origin of Mars' channels still remains to be discovered.

Thus we see that the landscapes of all the inner planets have been altered greatly over geological time by external agents. Planets lacking atmospheres now retain the heaviest scarring from asteroids and comets, which bombarded Earth and her sister planets early in their histories. Wind deposition and erosion have left their marks on the Earth and Mars and very probably on Venus. All the inner planets have been modified by gravity pull. Despite the vital new insights provided by spacecraft into the evolution and composition of the planets and the Moon, many important questions, such as the origin of the martian channels, are still unsatisfactorily answered and remain as the challenge for future scientific exploration.

SUGGESTED READING

Baker, V. R. "The Spokane Flood Controversy and the Martian Outflow Channels." *Science* **202**:1249–1256, 1978.

Bagnold, R. A. *The Physics of Blown Sand and Desert Dunes.* London: Methuen and Co., Ltd., 1954.

French, B. V., and N. M. Short (editors). *Shock Metamorphism of Natural Materials.* Baltimore: Mans Book Corp., 1968.

Hartmann, W. K. "Cratering in the Solar System." *Sci. Am.* **236**(1): 84–99 (Jan. 1977). (Offprint No. 351.)

Iversen, J. D., R. Greeley, and J. B. Pollack. "Windblown Dust on Earth, Mars, and Venus." *J. Atmos. Sci.* **33**:2425–2429, 1976.

Moorbath, S. "The Oldest Rocks and the Growth of Continents." *Sci. Am.* **236**(3):92–104 (March 1977). (Offprint No. 357.)

Roddy, D. J., R. O. Pepin, and R. B. Merrill (editors). *Impact and Explosion Cratering.* New York: Pergamon Press, 1977.

Sharp, R. P., and M. C. Malin. "Channels on Mars." *Bull. Geol. Soc. Am.* **86**:593–609, 1975.

Sharpe, C. F. S. *Landslides and Related Phenomena.* New York: Cooper Square Publishers, 1968.

4

RENEWAL
FROM WITHIN

RENEWAL FROM WITHIN

RESTLESS PLANETS

The production, transfer, and loss of heat is the fundamental internal source of modification of the surfaces of planets. Great amounts of heat were generated by large impacts on the surfaces of the Earth and the other inner planets during their formation, and probably, as well, by the decay of short-lived radioactive elements. Many scientists now believe the planets were completely molten at birth. Much of the heat left over from formation has flowed out of at least the mantles of the inner planets to their surfaces, whence it was lost eventually to space by radiation. However, additional heat has been continuously created through the gradual decay of long-lived radioactive material, especially uranium, thorium, and potassium.

As a consequence of initial and continued heating, materials deep within the planets remained molten over long periods of geological time. The outward flow of internal heat in turn caused deformation of all the inner planets' surfaces in response to differential movements of the molten material. On Earth, at least, this process continues actively to the present; on the Moon and Mercury, the outer layers cooled sufficiently billions of years ago to isolate their surfaces from internal deformation.

A major form of surface modification of the inner planets is the eruption of molten rock onto their surfaces. In fact, volcanism seems to be ubiquitous in the inner Solar System and has been discovered recently on the satellites of Jupiter by Voyager 1. Grabens (downdropped, linear structures) are another form of surface deformation that apparently is displayed on all the inner planets, indicative of crustal extension. Earth and probably Mercury have also undergone compression of their crustal layers. In addition, the surface of the Earth has experienced large-scale horizontal shear along regions of the crust that have been displaced laterally.

What Governs the Shape of Volcanic Landforms on Earth?

If we are to use the form of volcanic features as a clue to their origin on other planets, then we must understand volcanism as it is displayed on Earth. Everyone has seen pictures of various volcanic features, which range in shape from tall conical mountains to broad, flat plains. Why do these features take the shapes that they do?

The form of a volcanic feature depends primarily on the composition of the erupted magma and on how it is released at the surface. Other factors include the rate of extrusion of the magma and other volcanic products, the type of vent that releases the material, and the proportions of liquids (lava), solids (pyroclastics), and volatiles.

The viscosity (resistance to flow) of a magma is of paramount importance. Relatively nonviscous, or fluid, lavas (like iron-rich basalts) tend to flow long distances, even over gently sloping land. On the other hand, more viscous lavas (usually with higher silica content) tend to form steeply sloped, dome-shaped volcanoes, such as California's Mt. Lassen. A magma's viscosity is largely determined by its composition, temperature, and content of volatiles. Magmas containing abundant volatiles in solution can be quite fluid, but as soon as the volatiles exsolve and form bubbles (which become the vesicles in the hard rock), the viscosity can increase significantly. Highly volatile lavas often erupt explosively, creating craters of all sizes. Thus the morphology of volcanoes is strongly influenced by the amount and state of volatiles in the lava as well as the proportion of iron to silicon. In general, increased viscosity and volatility result in greater topographic expression. Basalts, the dominant volcanic rock on Earth, and probably on all the other inner planets as well, are iron rich and generally of low volatile content, leading to relatively low relief, but extensive volcanic terrains.

The size and shape of the vent have an influence on the morphology of volcanic deposits. If extrusion is confined to a single localized area, a large volcanic construct can develop. The island of Hawaii is made up of five such basaltic constructs that have developed over several million years: Mauna Kea, Mauna Loa, Kilauea, Hualalai, and Kohala volcanoes. However, if magma is erupted simultaneously at high rates from linear fissures over a large area, extensive lava plains may be created, as on the basaltic Columbia River Plateau. Such plains often obscure the actual fissures.

Volcanic Features on Other Planets

Differences in atmospheric conditions markedly influence the appearance of volcanic extrusions. The Moon and Mercury have no atmospheres. Therefore, the emplacement of pyroclastic deposits would be expected to differ from those on Earth because of the lack of atmospheric drag. For example, cinder cones, which on Earth have slopes up to 30° would, if formed on the Moon, have slopes of only one or two degrees. Cinder cones are made when lava droplets are shot into the air and then cool, solidify, and descend as "cinders." Because the Moon has no atmospheric drag, the cinders would travel for a longer time and spread farther. Thus the rim height of a lunar cinder cone would be less than one-tenth that on Earth, and the diameter would be at least four times greater. Some dark-haloed craters (Fig. 4.1) on the Moon are now interpreted to be basaltic cinder cones.

Figure 4.1
Oblique photograph taken by the Apollo 16 astronauts southward over the 117-kilometer-diameter lunar crater Alphonsus showing several elongated dark-halo craters on the floor of the crater. Most of the craters are associated with fractures and are considered to be volcanic. Alphonsus is a pre-Imbrium crater that has been modified by ejecta from the impact that formed the Imbrium Basin.

Another major environmental difference between the inner planets is the effect of surface gravity on volcanic phenomena. The very large, sinuous lunar rilles have been interpreted by some as collapsed lava tubes formed on a much larger scale than is possible on the Earth (Fig. 4.2) because of the Moon's low gravity—only one-sixth that of Earth's. Although detailed modeling has not been conclusive, these larger tubes presumably gain structural support by virtue of their much lower weight.

Similarly, on Mars—with about one-third the Earth's surface gravity—the calderas atop the very large volcanoes are much larger than similar features on Earth. It is possible that the lower martian gravity permits development of larger collapse structures than on Earth.

In interpreting volcanism on other planets, one must of course rely heavily on circumstantial evidence. On Mars, for example, rimless craters located on the flanks of what is obviously a very large volcanic construct are suspected of being of volcanic origin themselves. Similarly, chains of rimless craters oriented parallel to linear structural features on the Moon as well as on Mars are more likely to be of internal origin than to have formed by a chance alignment of uncorrelated impacts. In some cases, the statistics of crater distributions may provide clues to nonimpact origin because the size/frequency of nonimpact craters may be discernibly different from the characteristic form of impact craters.

Lunar Orbiter photographs of the Moon taken in 1965–1966 showed flow fronts and other features diagnostic of volcanism. Detailed color mapping from Earth-based telescopic observation and determination of relative ages by superposition from photographs in the late 1960s showed that the lunar maria are of different ages and of possibly different compositions. Finally, detailed sampling of the Moon in the Apollo program demonstrated that, indeed, these color differences frequently correspond to different episodes (and compositions) of basaltic lava extrusion.

The plains of Mercury, photographed by Mariner 10, exhibit strong circumstantial evidence of volcanism, although there is no specific proof. It is argued that the very large volume of material that fills basins there could not have come from the impact that formed those basins themselves, hence the material

Figure 4.2

Lunar Hadley Rille, region of the Apollo 15 landing, is a lunar sinuous rille more than 115 kilometers long that apparently originated from the elongate cleft at the bottom of the picture. This rille, like most sinuous rilles on the Moon, is believed to be a lava channel, parts of which were roofed during flow activity to produce a lava tube. The proportion of roofed to unroofed segments varies from rille to rille. These enormous flow features are many times larger than their terrestrial counterparts and are believed to play an important role in the emplacement of basalts on the Moon. Also visible on the left and top are several grabens (troughlike depressions between parallel faults), some of which have been buried by younger mare lava flows.

is inferred to be of volcanic origin. Plains units with differences in the degree of cratering indicate a difference in age and therefore require a mode of formation that operated over a long period. A volcanic origin best satisfies these constraints. Nevertheless, until truly diagnostic morphological features clearly identifiable with an eruptive process are recognized in future photographs of Mercury, the possibility of impact origin for some or all the mercurian plains cannot be entirely ruled out.

Some of the martian plains units are probably of eolian (wind) origin, and not volcanic at all. The difficulty again is one of finding diagnostic volcanic features to establish the origin. However, in portions of Mars the plains clearly do exhibit flow fronts and other features strongly suggestive of volcanic origin. The Viking Lander results suggest basaltic lava rock cropping out at the surface in one site and probably volcanic boulders at the other. Still, the origin of older plains, as on Mercury, is uncertain.

Chemistry and Petrology of Basalts

What are the composition and origin of the basaltic rocks that play such significant roles on the surfaces of the inner planets? We have actual samples only from the surface of the Earth and the Moon.

Terrestrial basalts have crystalline textures ranging from fine- to coarse-grained, depending partly on how slowly the lava cooled. Mineralogically, they are composd of plagioclase feldspars and ferromagnesian silicates (e.g., olivine and pyroxene), the latter giving basalts their characteristic dark appearance, plus many accessory minerals.

Seismic and geochemical studies indicate that basaltic magma on Earth is generated in the lower-crust/upper-mantle zone, at depths of 50 to 300 kilometers. Basaltic magma may erupt directly from its zone of origin or be shifted to a reservoir— a sort of "holding tank"—and then extruded later. Magma reservoirs and zones of origin are poorly understood, as is the "plumbing" that feeds lavas to the surface.

Some of the 382 kilograms of lunar rock that have been brought to Earth consist of fragments of crystalline basalts from lunar maria. In many respects these basalts resemble those of Earth; they are gray, fine- to coarse-grained crystalline silicate rocks, rich in ferromagnesian minerals. They are often vesicular (full of holes caused by gas bubbles), sometimes glassy, and much older than originally expected. Because of their relatively uncratered surfaces, lunar maria were once considered to be comparatively young. Yet radiometric ages of some of the younger mare lavas show them to have crystallized more than 3 billion years ago. Lunar mare volcanism took place from at least 3.8 billion years ago to possibly 2.5 billion years ago. It was also found that volatile materials—water, carbon, nitrogen, sulfur—are depleted or absent on the Moon, whereas refractory elements (those that are not easily melted), such as titanium, are relatively abundant. These differences from Earth basalts suggest that some higher-temperature episode took place very early in the formation of the Moon.

Although there are many ways to classify the lunar basalts, one particularly useful way for the mapping of mare lava flows is by titanium content. An early titanium-rich mare phase flooded large areas of the eastern part of the lunar nearside (sampled by Apollo 11 and 17 landings). This was followed by a less titanium-rich phase that flooded widespread areas (represented by Apollo 12 and 15 basalts) and by a second titanium-rich phase that flooded parts of Mare Imbrium and Oceanus Procellarum.

Laboratory simulations and physical-chemical data suggest that the basalts were produced by parent magmas generated at depths ranging from 100 to 500 kilometers, with the high-titanium lavas derived at shallow depth and the low-titanium lavas from deeper regions.

Studies of the viscosities of lunar lavas show that they were extremely fluid, about the same as motor oil at room temperature! Such lavas would tend to be erupted at very high rates and spread over vast areas—an expectation consistent with photographic evidence of very long individual lava flows on the lunar surface (Fig. 4.3).

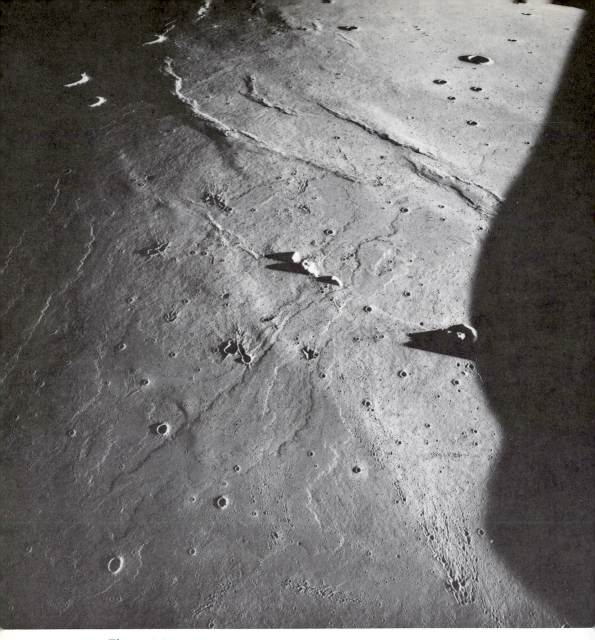

Figure 4.3
These lava flows in Mare Imbrium (lower left) are very flat and nearly level surfaces. Their apparent thinness indicates that the lava must have been of very low viscosity and was erupted at a rather high rate.

PLATE TECTONICS

For the Earth, the major patterns of volcanism are best explained by the widely accepted theory of plate tectonics. According to this idea—for which there is a growing mass of
evidence—the crust is composed of some fifteen major plates
that are "floating" on a plastic layer. Volcanism and earthquakes are largely confined to the borders of these plates
(Fig. 4.4).

The great plates are continually being generated at the mid-
ocean ridges, and are translated horizontally away from them.
New basalt upwells at these ridges and then cools, forming
crust that spreads gradually away from the ridges, usually for
thousands of kilometers, and eventually plunges into deep
"trenches," such as the Aleutian Trench off the Alaskan coast.

Observations of the sea floor show that: (1) rocks increase in
age with distance from the ridges, (2) sediments become
thicker with distance from the ridges, and (3) magnetized minerals in oceanic crustal materials record temporal variations in
the terrestrial magnetic field. These field reversals produce
magnetic "stripes" on the ocean floor that are symmetric about
the ridges and progress up to the trenches and there disappear.
Regions of crustal formaton are called spreading centers; regions of crustal destruction are called subduction zones.

Plate Boundaries—Where the Action Is

With the new recognition that tectonic activity is connected
over enormous distances, many previously isolated geologic
features have taken on new meaning. The boundaries of crustal
plates are recognized as the regions of greatest geological activity. Three main types of plate boundaries—and the phenomena
associated with them—govern much of the volcanism and deformation of the Earth's surface: spreading centers, subduction
zones, and transform faults.

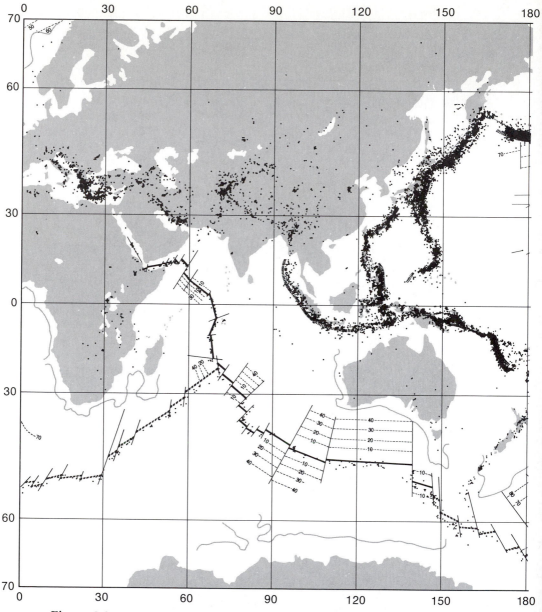

Figure 4.4
Worldwide pattern of seafloor spreading is evident when magnetic and seismic data are combined. Mid-ocean ridges (heavy black lines) are offset by transverse fracture zones (thin lines). On the basis of spreading rates determined from magnetic field measurements, isochrons have been established that give the age of the sea floor in millions of

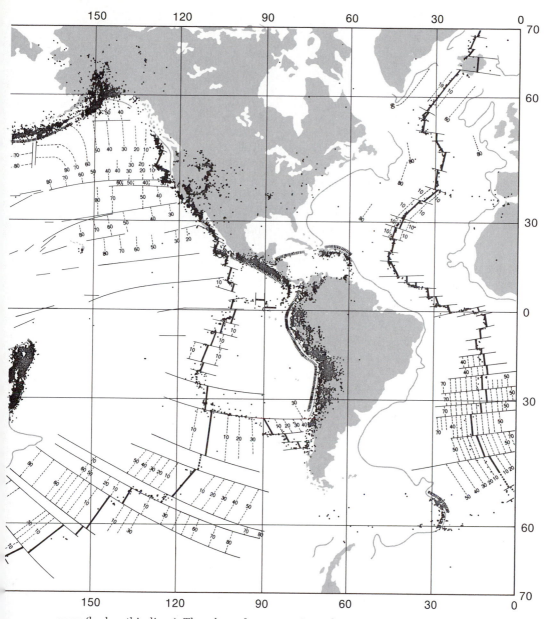

years (broken thin lines). The edges of many continental masses (gray lines) are rimmed by deep ocean trenches (hatching). When the epicenters of all earthquakes recorded from 1957 to 1967 are superimposed (dots), the vast majority of them fall along mid-ocean ridges or along the trenches, where the moving sea floor turns down. [From "Seafloor Spreading" by J. R. Heirtzler. Copyright © 1968 by Scientific American, Inc. All rights reserved.]

Most spreading centers are along mid-ocean ridges, which are characterized by rugged highlands, often rising 2 kilometers above the ocean floor, and exhibiting a higher-than-average flow of heat to the surface. Incised in the crest of most ridges is a deep cleft called a median valley, or rift (Fig. 4.5). Volcanic mounts are prevalent. Where ridges make abrupt bends, they are segmented into many small parallel offset sections separated by arcuate and roughly concentric fracture zones.

Although most spreading centers are beneath the ocean, in two prominent areas new crust is being formed above sea level. Iceland lies atop the Mid-Atlantic Ridge and is one of the most volcanically active regions in the world. A zone of ridges and faults splits the island and is the site of voluminous emissions of lava (Fig. 4.6). A second region is in East Africa, where spreading centers within the Red Sea and Gulf of Aden join and continue south through Ethiopia, Sudan, Uganda, Kenya, and Tanzania. The rate of crustal spreading for these areas is slower than on the ocean floor, where it can range from 1 to 10 centimeters a year.

In order for the surface area of the Earth to remain constant, old crust must be destroyed at the boundaries of colliding plates at the same rate as new crust is formed at the spreading centers. The geological, geophysical, and geographical configurations of the ocean trenches confirm that these occur in conjunction with plate boundaries and are indeed sites of crustal destruction (Fig. 4.7). As two plates collide, one descends beneath the other (subduction). In the simplest case, oceanic crust of high density plunges beneath continental crust of low density. As the subducted crust dives deep into the mantle, it is reheated and partially melted, generating new continental material of higher silica content than oceanic crust. Huge mountain ranges of igneous rocks, including volcanics, adjoin such subduction zones. Parts of the South American Andes are examples of this process (Fig. 4.8). In the case of two continental plates colliding, they scrape up ocean-floor sediments between them, folding and faulting huge tracts of land. The Alps are the

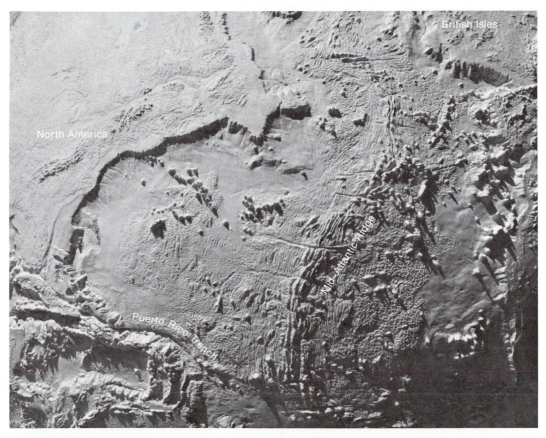

Figure 4.5
View of the North Atlantic Basin showing the Mid-Atlantic Ridge, a major crustal spreading zone, and associated fractures. This photograph is of an unpainted plaster model ("Lithosphere") having 25X vertical exaggeration. (Photograph of the Lithosphere, David A. Thomas.)

result of such a collision between Europe and Africa, and the Himalayas from the collision of India and Asia. In the third case of oceanic plate collison, oceanic crust is subducted beneath oceanic crust. The products of partial melting then migrate to the surface to create arcs of volcanic islands, such as the Philippines, behind the sea-floor trench. Subduction zones are the primary regions of mountain formation on Earth.

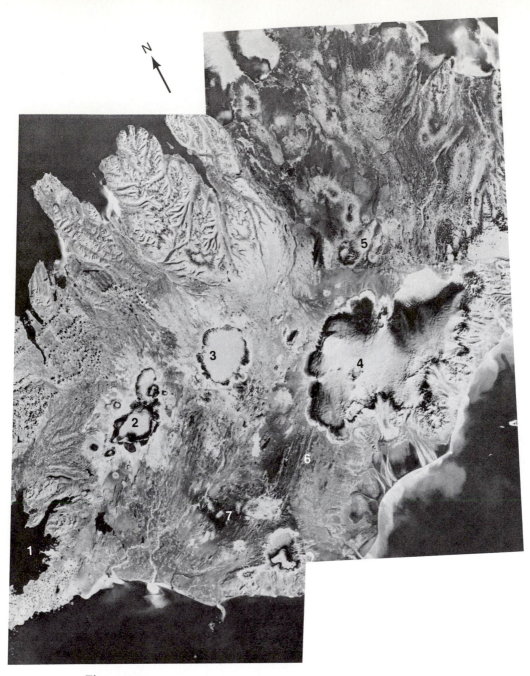

Figure 4.6
Landsat mosaic of most of Iceland. Iceland is astride the Mid-Atlantic Ridge and consists primarily of basaltic lavas associated with the spreading plates. Features identified are: (1) Reykjavik, the capital, (2) Langjökull (glacier), (3) Hofsjökull (glacier), (4) Vatnajökull (glacier), (5) Askja volcanic center, (6) Laki eruptive center, and (7) Hekla volcano. Distance across mosaic from lower left to upper right is about 400 kilometers.

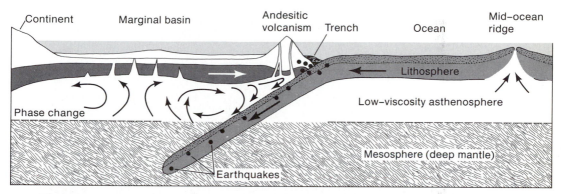

Figure 4.7

Formation and subduction of the lithosphere are shown in this cross section of the crust and mantle. New lithosphere is created at a mid-ocean ridge. A trench forms where the lithospheric slab descends into the mantle. Earthquakes (solid squares) occur predominantly in the upper portion of the descending slab. Arrows in soft asthenosphere indicate direction of possible convection motions. Secondary convection currents in asthenosphere may form small spreading centers under marginal basins. [From "Subduction of the Lithosphere" by M. N. Toksöz. Copyright © 1975 by Scientific American, Inc. All rights reserved.]

Figure 4.8
Altaplano volcanics and the Eastern Andean Cordillera. This mosaic of two Landsat images is about 185 kilometers across and shows a region near the border of Bolivia, Chile, and Argentina. It illustrates the complex terrain features of continental margin mountain-building. In the right-hand portion of the mosaic are old sedimentary rocks that have been warped, followed, and faulted into topographic prominence during the active movement of the South American and Pacific plates that began over 200 million years ago. Most prominent are eroded, plunging anticlines in the upper half of the image. In the middle left-hand portion of the mosaic are light-toned sedimentary deposits derived during the past 5 to 10 million years from the Eastern Andean Cordillera to the right, the Western Andean Cordillera (not seen), and the Altaplano ("high plateau") volcanics, seen most prominently in the bottom third of the photograph. The circular feature at the bottom of the frame is a large (50-kilometer) pyroclastic sheet erupted from a central caldera now nearly completely filled by later viscous lava flows. Most of the prominent circular mountains nearby are large andesitic stratovolcanoes. Such volcanism always accompanies mountain-building during oceanic-oceanic and oceanic-continental plate collisions. The Andes are the result of oceanic-continental plate collision.

Transform faults are plate boundaries along which portions of crust slide laterally past one another rather than colliding. One example is the San Andreas fault (Fig. 4.9), a well-known transform fault, which formed when a subduction zone off the California coast ceased to accommodate the motion of the Pacific Plate, which was moving generally northward relative to the North American Plate. A similar feature is developing in China as the India-Asia collision begins to move by transform faulting toward the Pacific (Fig. 4.10).

Silicic Volcanism—Second-Generation Rock

About three-quarters of the Earth's crust consists of basaltic material, nearly all of it being part of the oceanic crust. Basalt is the primary type of igneous rock supplied to the surface from the mantle. Continental igneous materials, however, are produced as well by partial remelting of primary igneous and other rocks in subduction zones, often in the presence of such volatiles as water and carbon dioxide (Fig. 4.11). By virtue of their lower density, these chemically differentiated materials rise up within the higher-density rocks of the upper mantle.

Figure 4.9
Landsat image of southern California showing segments of the San Andreas fault, which forms the southwestern boundary of the Mojave Desert, and the Garlock fault. With a known length of more than 1600 kilometers, the San Andreas fault is one of the longest in North America. It is a right-lateral strike-slip transform fault (seaside crustal block moves northwest in relation to the other block).

Figure 4.10
The Altyn Tagh Fault, China. This fault is seen here in a Landsat view about 185 kilometers on a side. The Altyn Tagh is a left-lateral strike-slip transform fault. The land at the bottom half of the frame is moving right, toward the Pacific Ocean, in response to the forces exerted as India collides with Asia. The sense of motion is shown well in the left portion of the image, where at least three prominent drainage systems show bends associated with the relative movement across the fault.

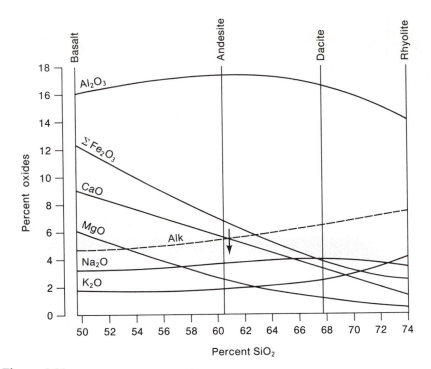

Figure 4.11
Volcanic rocks are classified according to their chemical composition. Shown here is a commonly used classification, Daly's average basalt-andesite-dacite-rhyolite variation diagram, in which different oxides are plotted against silica values. [From *Theoretical Petrology* by T. F. W. Barth, fig. III-60. John Wiley & Sons Inc., New York, 1962.]

Silicic lavas on Earth are usually (but not always) the by-product of plate tectonics. High-silica lavas of low volatile content produce thick, viscous flows that fragment easily as they are formed (Fig. 4.12). Commonly these lavas form volcanic domes, which vary greatly in size; some are only meters across, but others, like Mt. Lassen in California, are several kilometers in diameter and approach a kilometer in height. For lavas in which the amount of volatiles in the gas phase is larger, eruptions are more violent. Significant amounts of pyroclastic materials are explosively propelled from the volcanic vent. These

Figure 4.12
Oblique view of Big Glass Mountain, Medicine Lake Highlands, northern California. The very thick (note roads for scale) lava flows consist of blocks of obsidian and rhyolite several meters across and were emplaced as slow-moving viscous masses, typical of silicic lavas. Accumulations of these lavas contribute to the formation of volcanic domes.

eruptions are often characterized by their conical form; for example, Fujiyama in Japan, Mount Mayon in the Philippine Islands, Mount Vesuvius in Italy, and Mount Rainier in the state of Washington. These composite volcanoes, or stratovolcanoes, are formed by repeated eruptions of ash and several kinds of lava. Their slopes are characteristically steep and are composed of material ranging from the fluid basalts to viscous, highly silicic rhyolites. The lavas of Mount Etna in Sicily (Fig. 4.13) consist of basaltic and modestly silicic lavas and pyroclastic deposits. The slopes there are usually shallower than found on more silicic volcanic cones.

Figure 4.13
Skylab 3 photograph of Mt. Etna on the eastern coast of Sicily. This near infrared image shows numerous lava flows that make up the approximately 3600-meter-high volcano. Notice the numerous cinder cones that dot the upper flanks of the volcano. The small white streak at the summit is a fume cloud, indicating Mt. Etna's continuing activity.

Silicic volcanism can also produce, in contrast, huge expanses of flat-lying pyroclastic plains. Some of these plains are accumulations of *ash fall* (i.e., fine debris that settles from the sky after explosive volcanism), but most are formed by *ash flows*. Ash flows are superheated, fluidized mixtures of pyroclastic material and volatiles that travel more rapidly even than debris flows or landslides. The Valley of Ten Thousand Smokes in Alaska was filled by such deposits during the eruption of Mount Katmai. Deposits of this type usually become consolidated after emplacement by giving up their latent heat, and are then called *ignimbrites*.

Plate Tectonics on Other Planets?

What features of plate tectonics on Earth are visible from space? Most obvious are the continental and oceanic land masses, folded mountan ranges, mountain ranges of silicic volcanic material, island arcs, ridges of spreading crust (basaltic volcanic centers), and rift valleys.

Are these forms observed on other planets? Superficially similar landforms are seen on other planets, but their origins probably are different, except possibly for Venus. For example, the Valles Marineris complex of troughs and canyons on Mars (Fig. 4.14) bears a resemblance both in topographic and planimetric form to portions of the East African rift system. Yet it is unaccompanied by the smaller-scale volcanic features seen in Africa, although perhaps these could have been obscured by erosion. The large regions of tectonic features associated with Valles Marineris and its proximity to the Montes Tharsis volcanoes both suggest a tectonic origin. But the lack of compressional features (e.g., folded mountains) and the few, sparsely distributed conical volcanoes are suggestive of isolated magma sources (not subduction areas). Moreover, the lack of topographically defined continental and oceanic land masses strongly suggests that plate tectonics has not been operative on Mars. In addition, topographic and gravity data strongly suggest a rigid outer shell, much thicker than on Earth.

Mercury has a network of enormous escarpments that in places seem to have deformed craters by overthrusting. However, no spreading ridges, subduction zones, rifts, or folded

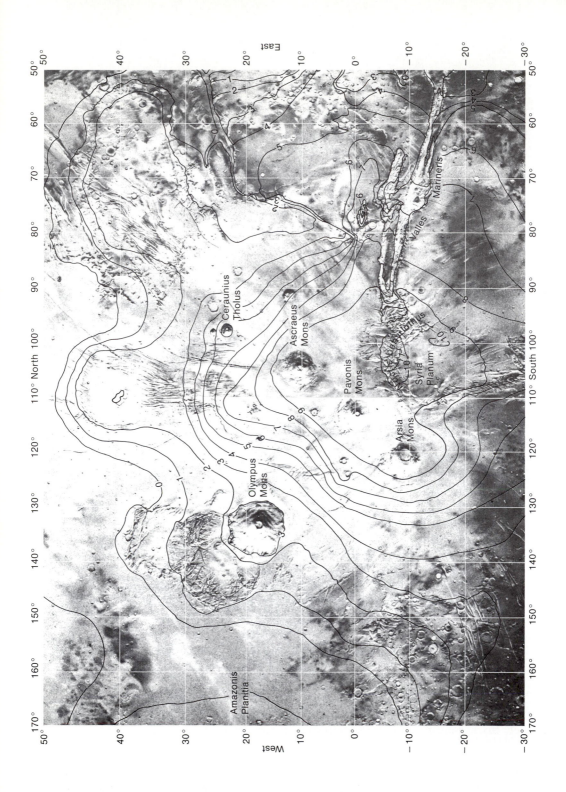

Figure 4.14

Tharsis region on Mars—the most prominent volcanic region on the planet—is shown in this shaded airbrush chart produced by the U.S. Geological Survey. The contour lines (omitted on the volcanoes) mark elevations above a reference level at which atmospheric pressure on Mars is 6.1 millibars. The contours outline the Syria Rise, a broad bulge in the martian crust some 5000 kilometers across and 7 kilometers high on which the shield volcanoes Ascraeus Mons, Pavonis Mons, and Arsia Mons are located. Fractures radiate outward for several thousand kilometers from the center of the bulge and were apparently created at the same time as the bulge. Valles Marineris may be part of the fracture system. [After "The Volcanoes of Mars" by M. H. Carr. Copyright © 1976 by Scientific American, Inc. All rights reserved.]

[149]

mountains are visible on the half of Mercury imaged by Mariner 10. Thus plate-tectonic activity seems unlikely there also. The Moon exhibits no large-scale compressional features at all. In addition, seismic data suggest a thick, rigid outer shell for the Moon, like Mars.

Radar observations of Venus offer some tantalizing views of that planet's surface, but at a resolution too low for detailed study. Features superficially resembling compressional mountains, isolated peaks arranged in arcs in close association with mountainous terrain, arcuate ridges, a large trough (Fig. 4.15), and continental-scale topographic variations suggest the possibility of plate-tectonic activity. In addition, the high surface temperature and Earthlike density and size imply molten conditions much nearer the surface than Mars, Mercury, or Moon.

Why don't Mars, Mercury, and the Moon exhibit plate-tectonic features? One possible answer is that the high internal temperature required to develop mantle convection, which moves Earth's plates, has been entirely dissipated because of the smaller size of those planets (greater surface area to volume ratio). Only Venus is comparable in size to the Earth, and on the basis of size scaling, its surface might plausibly exhibit plate tectonics. Another *ad hoc* possibility is that the original crust of the Earth simply may have been thinner than that on the other planets and hence easier to rift apart. Finally, the interior of the other planets may lack regions beneath their crusts where the right combination of temperature, pressure, and composition produce a zone of lowered friction upon which crustal plates can glide. The eventual determination of the presence or absence on Venus of topography characteristic of plate tectonics (presumably by a radar-equipped imaging satellite) will be of great significance to understanding Earth by providing insight into which global parameters really are crucial to the development and operation of plate tectonics.

Figure 4.15
Venus Trough. **Top.** This mosaic of two Earth-based radar reflectivity images covers an area some 2600 by 1700 kilometers. Note the large, troughlike feature (T) in the left image and the mottled plateau (P) in the right image. Note, too, the small canyonlike features (C) that cut the plateau. Image resolution is about 20 kilometers. The trough can be interpreted as possible evidence of crustal rifting. **Bottom.** A mosaic of two radar altimetry images of the same area; bright areas are topographically high, dark are low.

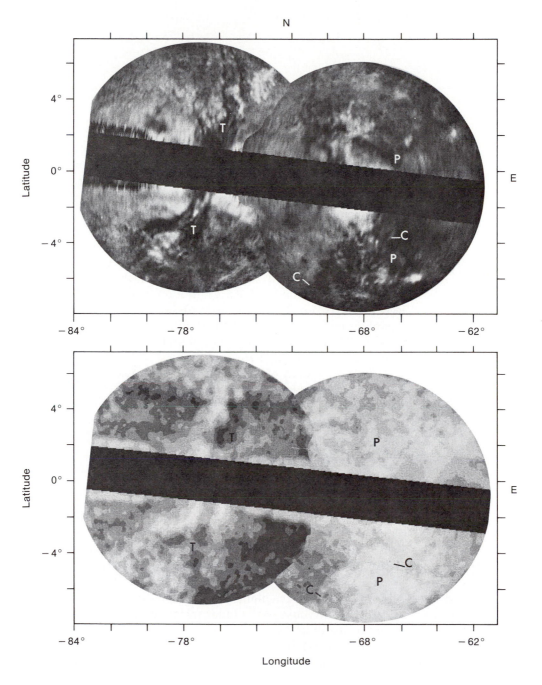

VOLCANIC PLAINS

All of the Earthlike planets have extensive plains. Those plains that appear to be of volcanic origin can be used to help us understand the surface and the internal thermal histories of planets.

On Earth, basaltic lavas can be erupted at very high rates, resulting in the spread of large volumes of lava over hundreds of square kilometers. Before sea-floor exploration began, the largest known basaltic regions on Earth were the continental plateau areas, such as the Columbia River Plateau and the Deccan Plateau of India. As more was learned about the ocean basins, it became apparent that basaltic lavas floor most of the oceans. Taken together, continental and oceanic basalts constitute most of the Earth's surface.

About 17 percent of the lunar surface is composed of mare basalts, which are restricted almost entirely to the side of the Moon that continually faces the Earth. The thinner nearside crust allowed internally produced lavas to surface more readily than did the thicker farside crust; those lavas now fill relatively low-lying regions of the Moon, typically the large impact basins, although some mare plains also occur outside of recognizable basins. Radiometric ages of lunar samples show that lavas filled the major impact basins several hundred million years *after* the formation of the basins, a relation of considerable significance in interpreting the origin of the lavas. Obviously, those lavas cannot be merely the "melt" left over from the impacts, but are true volcanic rocks originating deep within the Moon.

Mars also has vast plains units with some surface features very similar to those of the lunar maria (Fig. 4.16). Plains units occur in many areas, such as Lunae Planum; on the floors of some of the martian basins, such as Hellas; and in much of the northern hemisphere between the cratered uplands and the

Figure 4.16
In this moderate-resolution Viking Orbiter image of Mars showing multiple floodlike lava flows south of Arsia Mons, the lavas partly bury the heavily cratered terrain of the southern hemisphere, near the crater Pickering. The lavas flowed generally toward the south (toward the right in this view), partly surrounding the impact crater and flowing in the grabens in the middle of the view. Area of photograph is about 109 kilometers by 144 kilometers.

polar units. High-resolution photographs acquired from the Viking Orbiters show that many martian smooth plains were emplaced, like their lunar counterparts, by repeated outpourings of lava from linear vents that were covered in the process.

Of more uncertain origin than the martian plains are the plains of Mercury. Analysis of the morphology, distribution, and apparent age relation of these plains suggests that they are of volcanic origin, although a nonvolcanic interpretation has also been proposed.

Even though there are uncertainties as to the origin of plains on Mercury and Venus, basaltic lavas do appear to make up a significant part of the surfaces of all the inner planets.

Basalt Surface Features—Clues to Eruptive Styles

Studies of active volcanoes on Earth show that their morphology is the result of many complex and often interrelated parameters. The approach often taken in planetology is to use surface morphology to interpret the processes involved in the evolution of the surface. In this section, we contrast the morphology of two types of basaltic surfaces—flood-basalt plateaus and plains basalts (Figs. 4.17, 4.18). A higher-relief form of basalt volcanism, the shield volcano, is discussed at length later in this chapter.

Flood-basalt plateaus, typified on Earth by the Columbia River Plateau, are built up from individual flows that are commonly more than 100 kilometers long. The resulting flows typically are ponded as large lava lakes that bury not only the underlying topography but the feeding fissure vents as well. Consequently, it is often difficult to identify source areas. Such flow features as lava tubes and lava channels (Figs. 4.18, 4.19, 4.20) were either destroyed or were never developed in the first place. In some cases an "imprint" of the underlying topography may show through the surface of the flood basalt, possibly as a result of differential settling as the thick flows solidify. Otherwise, virtually no surface features are preserved on flood basalts, at least not on a scale that would be observable in spacecraft photographs.

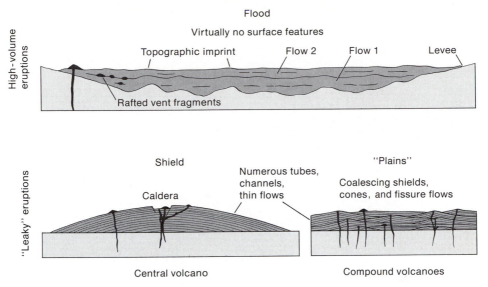

Figure 4.17
This diagram illustrates the distinction between flood basalt plateaus (high-volume eruptions from fissures), shield volcanoes (lower-volume eruptions from central vents and associated rift zones), and basalt plains (moderate-volume eruptions from both central vents and rift zones).

Figure 4.18
A typical basaltic lava plain consists of multiple coalescing low shields, buried shields, and fissure flows. The relatively thin flows are usually fed by tubes or channels.

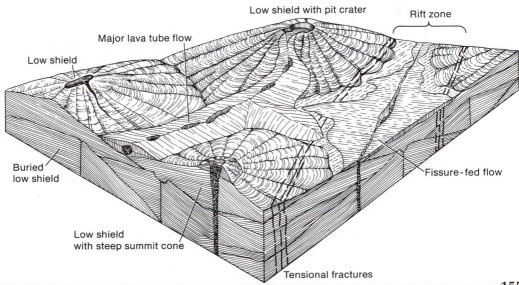

lateral levee

Figure 4.19
A leveed channel on Earth. This small lava channel with lateral levees developed in a thin flow on the flanks of Mauna Loa shield volcano, Hawaii.

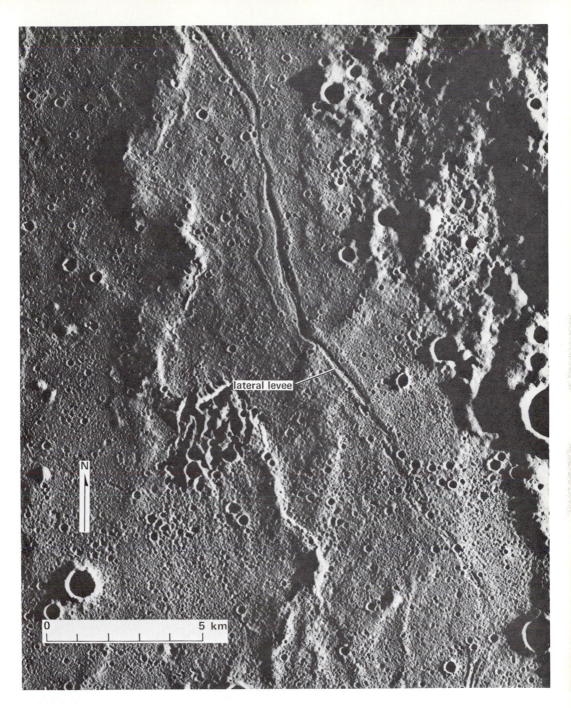

lateral levee

Figure 4.20
A leveed channel on the Moon. Photographed during the Apollo 16 mission under low-angle illumination, this lunar sinuous rille west of Bonpland F crater shows lateral levees and the braided character typical of smaller rilles.

In contrast, basaltic plains, such as the Snake River Plain in Idaho, are characterized by lower volumes of lava per eruption than flood basalts, and are constructed of flows that typically are 10 meters or less thick. Vent structures (Fig. 4.21, 4.22) are usually preserved, and flow features are common. Advancing flows typically are fed by lava tubes and channels, which often repeatedly serve as conduits for multiple eruptions and can be enlarged by erosion of earlier-formed lavas.

The plains are composed of numerous, small, circular, coalescing *low shield volcanoes* with very low profiles (slopes typically less than 1° or 2°) and intervening fissure flows. Many low shields have summit craters and are aligned on rift zones.

Surface features on lunar maria include mare ridges (Fig. 4.23), sinuous rilles, domes (Fig. 4.24), small shieldlike

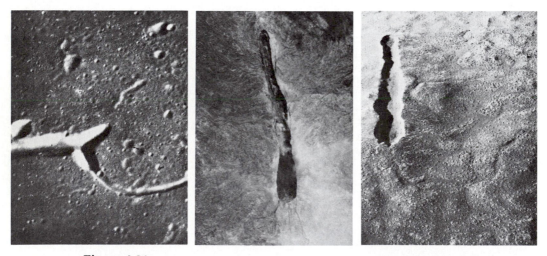

Figure 4.21
Elongate vents common on basaltic plains on the Moon (left) are compared with those on Earth (right; Snake River Plain, Idaho). During the growth of the summit vent at Mauna Ulu, Hawaii (middle), the formation of the elongate vent was seen to grow from a series of individual vents oriented on a fissure into a single structure. Lava spilled from the vent by overflow as well as through lava tubes and channels—flow features that are seen on the lunar analog as well as in the Snake River Plain.

RENEWAL FROM WITHIN

constructs, and depressions with irregular planimetric forms—
all probable volcanic features. Lunar sinuous rilles appear to be
huge collapsed lava tubes or lava channels that fed mare basalts
into the basins. Some mare ridges may be deformational features that developed on the crust of vast lava lakes during solidification. Others could be volcanic vents of lavas more viscous than the typical mare basalts; still others are probably true
tectonic features associated with minor faulting.

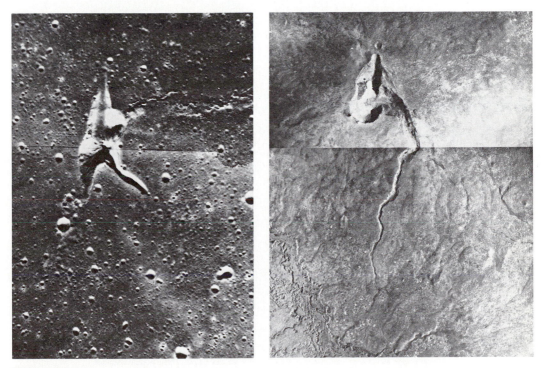

Figure 4.22
Two structures having similar morphology on Earth (right; Bear Crater) and Moon (left;
Aratus CA). Bear Crater is a 780- by 254-meter basaltic vent on the Snake River Plain,
Idaho, that has a 3.5-kilometer long "tail"—a lava channel. Aratus CA is a 9.5 kilometer-
long depression in western Mare Serenitatis that is interpreted to be a small vent. It has a
small sinuous rille trending away from the crater. Both Aratus CA and Bear Crater display multiple eruptive phases. Although these two features differ in size by a factor of 10,
the similarity of morphology and geological setting (i.e., basaltic plains) enhances the
analogy.

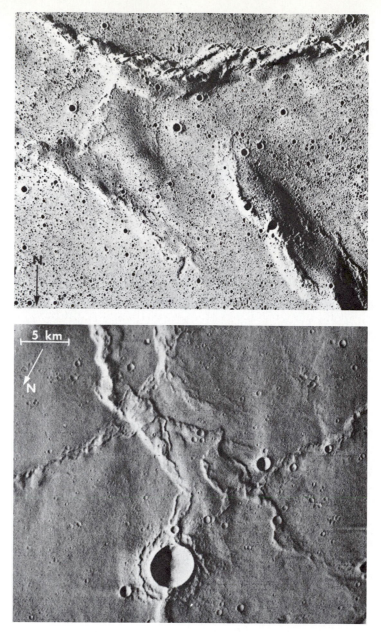

Figure 4.23
Mare ridges on the Moon (top) are compared to ridges on martian smooth plains (bottom). The lunar maria, formed by enormous eruptions of basaltic lava from linear vents, often display ridges around their margins. These ridges are believed to arise as the lunar interior adjusts to the mass of the lavas. The general and detailed similarity of the martian ridges suggests comparable origins. The lunar example is from Mare Serenitatis (near 19.5°N, 20.3°E) and was taken when the Sun was but 5° above the horizon. The Mars example is from Ter Planitia (near 23.2°S, 244°W) and was taken at a Sun elevation angle of 20°. Both pictures are reproduced here at the same scale.

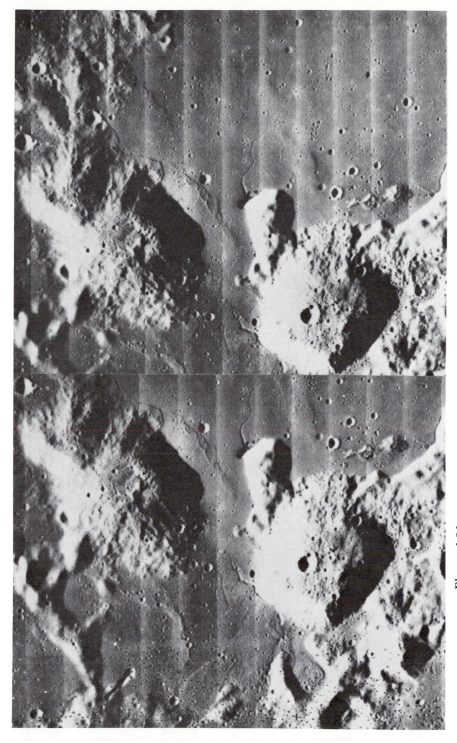

Figure 4.24

This stereoscopic view shows the Gruithuisen lunar domes, approximately 19 kilometers in diameter. These probable volcanic structures occur in several areas of the Moon and may represent extrusion of lavas more viscous than typical mare basalts.

On Mars, both flood-type and plains-type basalts appear to be present. As yet, it is not possible to correlate units with age and styles of emplacement on a planet-wide basis. Some areas, however, do show interesting correlations. The plains of Lunae Planum exhibit mare ridges, but no flow fronts, possibly indicating flood-type volcanism. On the other hand, Elysium and some Tharsis plains have some channels that may be collapsed lava tubes and flow fronts, but not mare ridges, possibly indicating plains-type volcanism.

Thus far, neither the careful examination of Mariner 10 pictures of Mercury nor preliminary geological mapping have uncovered any structures suggestive of plains-type basaltic volcanism. For example, no features can be identified positively as lava channels or collapsed lava tubes, endogenic domes, or constructional shield volcanoes. However, vast smooth areas have been found that are relatively uncratered and that have mare-type ridges and irregular fractures (Fig. 4.25). These areas, if volcanic, are probably flood-type lavas of basaltic composition.

When Is a "Plain" Not of Volcanic Origin?

Circumstantial evidence of volcanic origin was applied in part to a characteristic unit of the lunar surface that had been mapped first from telescopic observations and later from Lunar Orbiter photography prior to the Apollo missions. These relatively smooth plains of generally higher albedo than those of the mare regions are termed "light plains," and exhibit a somewhat higher abundance of impact craters than mare surfaces. They were generally interpreted to be a volcanic unit older than the mare plains that filled major lunar basins.

However, surface samples from one light-plains region, returned by the Apollo 16 mission, were discovered not to be of volcanic origin. They are a badly broken up composite of fragments—breccias—suggestive of some kind of impact rather than volcanic origin.

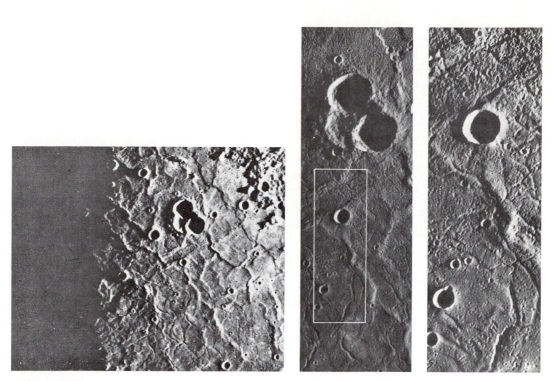

Figure 4.25
Shown here are ridges and fractures on the Caloris Plains of Mercury. The Mariner 10 frame on the right shows an area about 50 kilometers long from top to bottom.

The "light plains" have since been reinterpreted by some as locally derived debris churned up by secondary impacts. Alternatively, the plains may have been formed from an enormous blanket of fluidized ejecta created somehow during the event that formed the Orientale Basin, although orbital remote-sensing data do not display the consistent reflective signatures that would be expected if all the light-plains material were derived from a single gigantic impact event. The unexpected finding of nonvolcanic plains of the Moon, combined with the lack of definitive evidence of volcanic origin of the mercurian plains, requires that the mercurian question be kept open until future diagnostic data are acquired.

Lunar Mascons

In the mid-1960s, scientists noticed anomalous perturbations in the orbit of the Lunar Orbiter spacecraft. When the spacecraft passed over certain mare regions, they were pulled slightly closer to the Moon and increased in velocity. The reason was attributed to the gravitational attraction of concentrations of mass in certain regions of the Moon—hence the term "mascon." Subsequent analyses of spacecraft orbits resulted in the identification of positive lunar gravity anomalies for Mare Smythii, Mare Crisium, Mare Serenitatis, Mare Imbrium, Mare Nectaris, and Mare Humorum. All are circular impact basin sites.

One explanation for the mascons is that they are disk-shaped masses of dense basalt near the surface and/or an associated upwelling of higher-density mantle rock beneath the crust. In this interpretation, the basalt is presumed to have been emplaced during early flood-stage filling of the impact basins, when lava flows presumably were thick, often ponded, and typical of lava-lake activity.

In whatever manner the mascons formed, it is evident from the presence of positive gravity anomalies even today that the lunar crust has remained rigid enough to support such excess masses since their emplacement as much as 3.8 billion years ago.

Modifications of Pre-existing Impact Craters

Eruptions of basaltic lavas can affect pre-flow terrain in many ways. Most obvious is simple flooding and burial. Many times, however, even when the terrain may be covered with lavas to depths of tens of meters, the pre-flow terrain may continue to show through the lava flows. The "ghost" craters observed in many lunar maria and in some martian volcanic plains probably are impact craters that have been buried by basalt flows.

Large-scale volcanic modifications of impact craters may also take place. On Mars, Mercury, and particularly on the Moon, a great many large craters have floors that appear to have

Figure 4.26
This oblique southward view of the lunar crater Humboldt was taken by the Apollo 15 astronauts. The crater is more than 200 kilometers across and is of the type called "floor-fractured craters." The main characteristic of this class of crater is that the floor has a network of approximately radial and concentric fractures, which are interpreted to result from tension caused by intrusion of magma from below.

been extensively modified by volcanic and associated tectonic processes. The term "floor-fractured craters" has been applied to some of these geologic features. On the Moon, these modifications seem to have occurred during the emplacement of the mare lavas. Several different classes of floor-fractured craters (Fig. 4.26) have been identified on the Moon. The impact-

Figure 4.27 (*facing page*)
Mauna Loa (left) and Mauna Kea (right) shield volcanoes on the horizon (town of Hilo on the bay, right side). They exhibit a difference in profile that is attributed partly to the differences in lava composition and styles of eruption. Mauna Loa is built principally of basaltic lava flows, producing the typical, smoothly rounded, shield-shaped profile; Mauna Kea lavas have evolved to hawaiite (andesitic), and the style of eruption involves a higher proportion of cinder production. Most of the irregular knobs visible in the profile are cinder cones; the result is a steeper summit than on Mauna Loa.

brecciated zone beneath most such craters appears to have been intruded by lavas, and the floor, including the central peak, may have been elevated 100 meters or more. Lunar craters with central peaks higher than the rim are considered good evidence of such uplift.

SHIELD VOLCANOES

Although volcanic plains are by far the most extensive type of volcanic unit on the Earthlike planets, the features most people think of when volcanism is mentioned are cone-shaped strato-volcanoes, such as Mt. Hood or Fujiyama, and shield volcanoes, such as Mauna Loa in Hawaii (Fig. 4.27). The term "shield" was originally applied to volcanoes in Iceland for their resemblance to a knight's inverted shield. Now the term is applied to all volcanic edifices having a low, gentle profile, regardless of size. Very impressive shield volcanoes have been discovered on Mars, and small ones on the Moon. As for strato-volcanoes, they may be unique to plate tectonics and to the Earth.

Shield volcanoes on Earth range from a few kilometers to more than 100 kilometers across, and their total relief above the ocean floor approaches 10 kilometers (Fig. 4.28). Slopes of typical shield volcanoes on Earth range from 2° to 12°, with outer slopes merging imperceptibly with the surrounding terrain. At least in part, the general form of shield volcanoes is a function of the rate of eruption and volume of lava extruded. Lava erupted rapidly or in large volumes produces a flood-type flow; lava erupted slowly or in small volumes produces dome-shaped volcanoes with steep sides. Most shield volcanoes are formed by relatively fluid lavas (typically of basaltic composition)

Land volcanoes

Mt. Etna

Volcanic islands

Stromboli

6 km
0

0   10   20 km

Scale for all volcanoes
No vertical exaggeration

Caldera

Sea level

NE–SW across Mauna Loa, Hawaii

Figure 4.28
Differences in profile are readily apparent in this comparison (to scale) of terrestrial
shield volcanoes. Mt. Etna and Stromboli, composite volcanoes composed of basaltic and
andesitic lavas in the form of flows and tephra deposits, are more convex upward. In
contrast, Mauna Loa, the largest shield volcano on Earth, is composed of basaltic lava
flows distinctly more rounded in appearance.

erupted at fairly high rates either from a central vent or from localized vents along fissure systems; the volumes of lava are generally less than for flood-type eruptions and produce thin flows that frequently form lava tubes and channels.

The planimetric form of shield volcanoes depends on the character of the vent, local tectonic patterns, and adjacent structural features. For example, the shields of Iceland and the Galapagos Islands are rather circular, resulting from predominantly central vents. In contrast, the Hawaiian shields are elongated, a result of multiple vents located along regions of crustal fractures; each shield is constructed adjacent to an older shield.

Hawaiian Shield Volcanoes

The active shield volcanoes on the island of Hawaii are among the best known and most thoroughly studied volcanoes in the world. They have been monitored almost continuously throughout this century by volcanologists of the Hawaiian Volcano Observatory, located on the rim of Kilauea crater.

The Hawaiian Islands are the tops of volcanic edifices that rise from the sea floor along the Hawaiian Ridge. In all, more than 50 huge undersea volcanoes extend more than 3500 kilometers across the Central Pacific, making up the Hawaiian Archipelago. The islands of this chain become progressively younger toward the southeast, culminating in the island of Hawaii with its two active volcanoes, Kilauea and Mauna Loa.

The summits of the Hawaiian shield volcanoes are frequently marked with calderas—large, irregular craters (Fig. 4.29). The calderas typically result from collapse as the magma settles to lower levels or shifts to other locations. Repeated collapse at the summit causes a flattening of the top of the shield.

The magmas that form the Hawaiian shields tend to become slightly more silicic as the period of volcanism comes to a close. This results in the production of more pyroclastics and clinkery lava (called "aa"), which forms steeper slopes on the upper portions of the volcano.

Averaged over geologic time, the rates of eruption involved in the formation of oceanic shield volcanoes like Mauna Loa

Figure 4.29
The caldera of Kilauea volcano, Hawaii, shows the irregular concentric fracture pattern typical of calderas and a prominent pit crater, Halemaumau. Lava lakes typically rise and fall within the pit crater like a piston, occasionally overflowing the rim to flood the caldera floor. Lava may also erupt through fissures on the floor, as evidenced by the 1954 flow to the right of the pit crater. The caldera is about 4.3 kilometers long and 2.9 kilometers wide. Mauna Kea volcano is in the background.

are among the highest on Earth. Individual eruptions, however, have lower rates of extrusion and produce smaller total volumes of lava than individual eruptions of flood-type basalts. Put another way, volcanic activity that produces shield volcanoes is more frequent but produces less material per occurrence; eruptions of flood basalt are less frequent but produce much more material per occurrence. This may also explain why eruptions of flood basalts seldom recur from old vents, but instead always from new ones. There is enough time after each eruption for the erupted lava to solidify completely and to block the vent network that produced it.

Olympus Mons—A Giant of the Solar System

Among the more spectacular discoveries in the exploration of Earth's planetary neighbors are the huge shield volcanoes on Mars, first revealed by Mariner 9. Olympus Mons (Fig. 4.30) is the largest, tallest volcano on Mars; it measures some 500 kilometers by 600 kilometers across the base and stands more than 24 kilometers above the surrounding plain. In addition, lava flows originating on the volcano can be traced hundreds of kilometers beyond the prominent basal scarp of Olympus Mons onto the surrounding plains. Hawaii's Mauna Loa, the largest volcano and the largest single mountain on Earth, is only one-fifth the areal size and less than one-twentieth the volume of Olympus Mons. Another difference between the Hawaiian volcanoes and Olympus Mons is their shape. The Hawaiian shields tend to be elongate, but Olympus Mons is fairly circular, like the shields of Iceland and the Galapagos Islands, and does not display extensive rift zones, although some fissure-like features are observed on its lower slopes.

Otherwise, the general makeup of the martian shield is the same as on Earth—gentle slopes formed by thin, fluid lava flows, some of which exhibit tubes and channels (Fig. 4.31); a summit caldera displaying multiple collapse craters; and possible concentric fault systems. Olympus Mons has a multislope profile with steeper slopes on the upper parts of the shield. The steep slopes could reflect an evolution in eruptive style to a slightly more viscous lava or a shift to lower volumes of lava per eruption. The steeper part of the shield is made up of discontinuous terraces. Similar terraces are observed on the flanks of Mt. Etna.

The summit of Olympus Mons is marked by an enormous caldera more than 80 kilometers across. Large calderas on Earth result mainly from collapse. Detailed analysis of the summit of Olympus Mons suggests a similar mechanism.

Figure 4.30
Olympus Mons, the largest volcano in the known Solar System, was discovered on Mars during the Mariner 9 mission in 1971. Measuring about 500 kilometers by 600 kilometers, this enormous shield is composed of thousands of individual flows and flow units, analogous to the construction of its smaller counterparts on Earth. The upper half of the volcano has a distinctive terraced character.

Figure 4.31
This high-resolution picture of the flank of Olympus Mons shows thin, narrow lava flows and a prominent sinuous channel. These features are typical of terrestrial shield volcanoes and reflect the relatively fluid character of basaltic lavas. The sinuous channel is interpreted to be a collapsed lava tube. Note also the leveed channels in some of the narrow flows. Width of area shown is about 50 kilometers.

On Earth, collapse craters smaller than a kilometer or so in diameter are called pit craters. These rimless, steep-walled depressions typically occur on the flanks of shield volcanoes, on the summits associated with calderas, and, occasionally, on volcanic plains. Infrequently, pit craters may originate by explosive eruptions and be enlarged later by collapse. Some of the rimless craters near the summit of Olympus Mons and on other martian shield volcanoes may be pit craters.

Olympus Mons lies in a region of Mars that appears to have had a long history of volcanism. Most prominent are the volca-

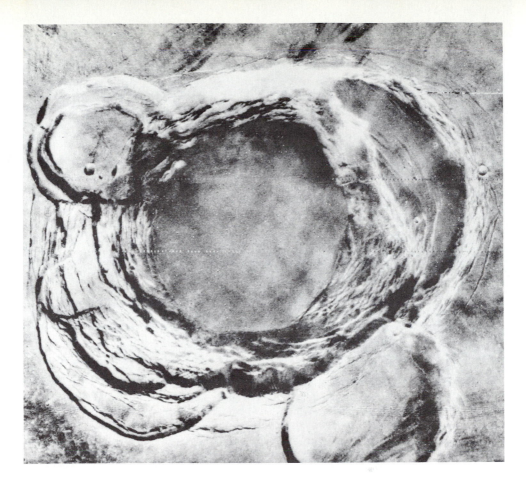

Figure 4.32
Moderate-resolution image of the 40-kilometer summit caldera of Ascraeus Mons, one of
the shield volcanoes of the Tharsis region of Mars. The complex concentric fractures,
scalloped rim, and multiple collapse craters are typical of calderas on Earth.

noes of the Tharsis Ridge (Fig. 4.14), an elevated area trending
northeast-southwest, which is part of a distinct bulge in the
martian crust known informally as the Syria Rise. The three
great shield volcanoes of Tharsis appear to be slightly different
versions of Olympus Mons (Fig. 4.32). Detailed examination of
the volcanoes show that, unlike Olympus Mons, most have a
riftlike zone oriented parallel with the ridge.

In addition to the enormous shield volcanoes, several other
volcanic edifices occur in the area, including small, dome-
shaped ones to the northeast (Fig. 4.33). These may have been
formed by more viscous lavas (possibly more silicic) or by

Figure 4.33
Tharsis Tholus, a volcanic dome about 120 kilometers across, is typical of several domes in the general area north and northeast of the Tharsis Ridge. The dome is fractured and has a prominent, complex summit caldera.

eruptions of small volume, either of which could produce a steep-sided construct. On some of the domes, small channels originate near the summit and extend radially down the flanks. If these are volcanic channels, then it is likely that the domes are composed of lavas only modestly more silicic than basalts. On Earth, more silicic lavas are too viscous to form lava channels.

Shield Volcanoes of the Elysium Region and Other Areas of Mars

A second concentration of shield volcanoes on Mars is found in the Elysium region—a broad, elliptical feature about 3000 kilometers long and 1500 kilometers wide that stands about 4 kilometers above the martian datum. Three large constructs (Elysium Mons, Albor Tholus, and Hecates Tholus) rise from this

Figure 4.34
This Mariner 9 mosaic shows part of Hecates Tholus and its summit caldera. The slightly sinuous, discontinuous depressions radiating from the caldera are partly collapsed lava tubes and channels, some aligned on probable rift zones. The small, bowl-shaped craters are probably impact features. Lava plains are superimposed on the west flanks (right side) of the volcano.

broad region. Thus, in some respects, the Elysium region is a small version of the Syria Rise. Most of the constructs do show radial fractures, collapsed lava tubes, lava channels, and complex calderas (Fig. 4.34). The plains on which the Elysium volcanoes occur also appear to be volcanic, but it is difficult to determine age relations between the plains and shields.

Comparison of the Elysium volcanoes with the shields of Tharsis and with terrestrial analogs (Fig. 4.35) suggests that some of the Elysium volcanoes may be composite cones—that is, interbedded lava flows and ash deposits. Morphologically,

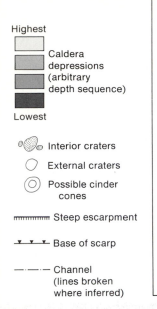

Highest

Caldera
depressions
(arbitrary
depth sequence)

Lowest

Interior craters

External craters

Possible cinder
cones

Steep escarpment

Base of scarp

Channel
(lines broken
where inferred)

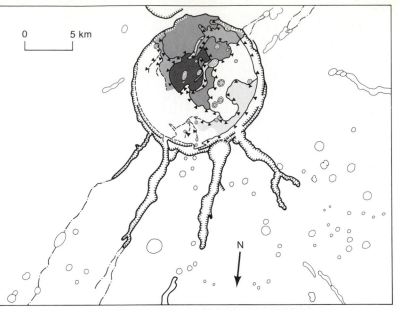

0   5 km

N

Elysium Mons, Mars

Elysium Mons, Mars

Figure 4.35
Left. Elysium Mons. (Top) Sketch map of Elysium Mons made from Mariner 9 photography. Individual flow units cannot be distinguished, but different elevation levels in the caldera may represent phases of activity. (Bottom) This Mariner 9 photograph of the

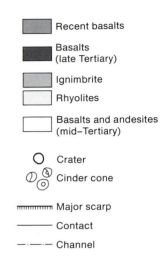

Recent basalts

Basalts
(late Tertiary)

Ignimbrite

Rhyolites

Basalts and andesites
(mid–Tertiary)

○ Crater

Cinder cone

Major scarp

Contact

Channel

Emi Koussi, Africa

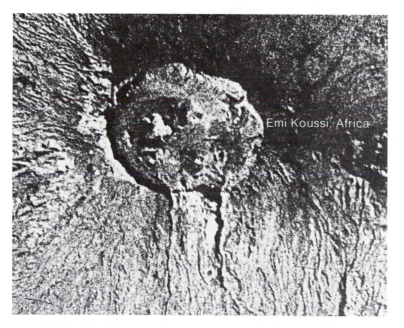

Emi Koussi, Africa

caldera shows subdued mounds and hummocky surface. **Right.** Sketch map (top) and photograph (bottom) of the summit region of the Emi Koussi volcanic complex in Africa. Compare with the map and image of Elysium Mons; summit calderas of both features have similar morphologies and are about the same size.

the hummocky texture of the flanks, with their smooth, possibly pyroclastic blankets and steep profiles support this possibility, which may imply volcanism of higher silica concentration in the Elysium region than elsewhere on Mars.

Several large volcano-tectonic features, termed "patera," have been discovered on Mars; these features appear to be unique to that planet. For example, Alba Patera is a low-profile feature, about 1600 kilometers across, identified by distinct radial patterns and concentric fractures. Some of the patera appear to be degraded, old shield volcanoes suggesting long-term volcanic activity on Mars.

Shield Volcanism on the Moon

The conspicuous basaltic flood-and-plains-type volcanism of the lunar maria tend to overshadow the presence of shield volcanoes on the Moon. Nonetheless, shieldlike volcanoes are present, although they are minor features by martian or terrestrial standards.

The shields are difficult to identify, which complicates the task of determining their number and significance. Those tentatively identified are small—typically less than 15 kilometers across—and have very low profiles (Fig. 4.36). They are best seen under low angles of illumination. Most often, lunar shieldlike features consist of a gentle swelling on the mare plains and have either a summit depression or a central "spike." Because their slopes are often less than about 5° and merge gradually with the surrounding plains, it is difficult sometimes to discern confidently whether many are true volcanic constructs or simply irregularities on the mare surface.

Other shieldlike features on the Moon are steeper-sided structures that might more properly be called "domes." Both the low-profile and the dome-like shields typically occur in areas that appear to be flooded with relatively shallow mare lavas, and seem to represent basaltic plains-type volcanism.

In conclusion, shield volcanoes are among the most impressive surface features observed on planetary surfaces. On Earth,

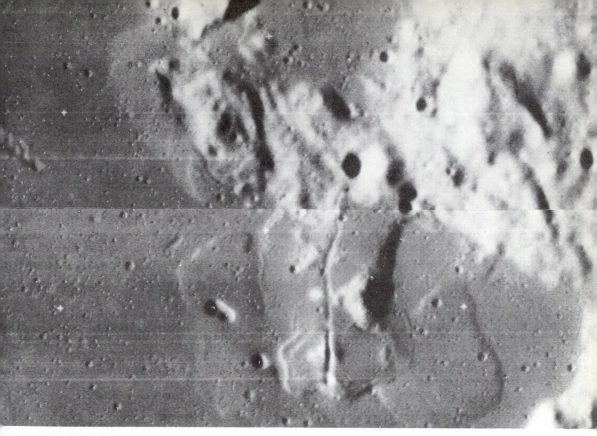

Figure 4.36
A small shieldlike lunar volcano about 12 kilometers in diameter. Irregular flow lobes, barely visible in this Lunar Orbiter IV picture, imply an extrusive volcanic origin. The linear fracture appears to be a fissure that fed some of the flows.

they occur largely on the ocean floor as parts of the giant submarine seamount chains. Because many are underwater features, they are not as obvious as continental mountains, but in fact are more voluminous and probably more active than most continental versions as well. Martian shield volcanoes are associated with crustal bulges so large that they alter the gross shape of the planet. Shield volcanism on the Moon plays a minor role, and is unrecognized on Mercury so far. The differences in the size of shield volcanoes and their possible absence on Mercury suggest significant differences in the characteristics of the crust and in the evolution of magma chambers from planet to planet.

CRUSTAL DEFORMATION ON THE EARTHLIKE PLANETS

The outer layers of all the Earthlike planets have cooled sufficiently since formation, by radiating their heat into space, that they have essentially been rigid for billions of years. At greater depth, however, the temperatures and pressures in at least some are still high enough to cause the constituent materials to act as viscous fluids. As a consequence, the relatively brittle skins of the Earthlike planets frequently manifest the blemishes of deformation arising from slow vertical and sometimes horizontal motions of the interior. On Earth, this process continues to the present time, with earthquakes being the most common manifestation of the deformation of the Earth's cool, rigid, outer layer (lithosphere). At least minor internal activity probably still continues on Mars, as suggested by the probable detection of a single marsquake by Viking Lander 2. On Mercury and the Moon, however, even minor vertical deformation generally appears to have ended billions of years ago. Answers about the present state and past history of surface deformation on Venus are almost within reach of U. S. and Soviet spacecraft.

The principal type of deformation on the Moon and Mars has been crustal extension. Portions of these surfaces have been literally pulled apart, yielding downdropped features such as grabens. On the Earth, there has been widespread compression as well. This has resulted in thrust faults, folded mountain belts, and the creation of high mountain ranges. Early in its history, Mercury experienced widespread crustal compression that led to a characteristic pattern of faulting still evident in ancient terrains. On Earth, in addition to crustal compression, shear has displaced large masses of crust laterally with respect to each other, as evidenced along the San Andreas and many other major earthquake faults. No unambiguous evidence of large-scale horizontal displacement has been recognized on the Moon, Mercury, or Mars.

Grabens

The most common deformational features of the Earthlike planets is a simple downdropped block called a graben. Grabens on Earth range from small features—a few tens of meters across and a few hundreds of meters long—to such enormous features as the East African rift zone, which extends over 3000 kilometers in a north-south direction and averages 50 to 100 kilometers in width. Grabenlike features were recognized on the Moon well before the space age, and were named linear and arcuate *rilles* (Fig. 5.10). Many lunar rilles cut mare plains that filled the lunar basins hundreds of millions of years after the basins were originally formed in cataclysmic impacts. In view of such an extended period between basin formation and rille development, these rilles probably cannot be explained simply as the consequences of subsidence associated with the formation of the basins themselves. Furthermore, in some places the rilles also have formed in highland terrain.

Complex graben patterns also are recognized on Mars, leading to application of the term "fractured plains" in some equatorial areas. There, grabens cross each other (Fig. 4.37), providing spectacular evidence of repeated episodes of crustal extension. The fractured plains are overlain by younger, smooth volcanic plains that do not evidence the same degree of fracturing. This may indicate that there have been discrete episodes of crustal extension in the planet's history. By way of contrast, the Mariner 10 pictures of Mercury seldom disclose grabens except for the high-resolution pictures of the Caloris Basin (Fig. 4.25), which do show small, grabenlike features. In other areas of the mercurian plains, grabens are rare. Thus the plains of Mercury show significantly less evidence of crustal extension than those of the Moon and Mars, suggesting corresponding differences in the history of its interior.

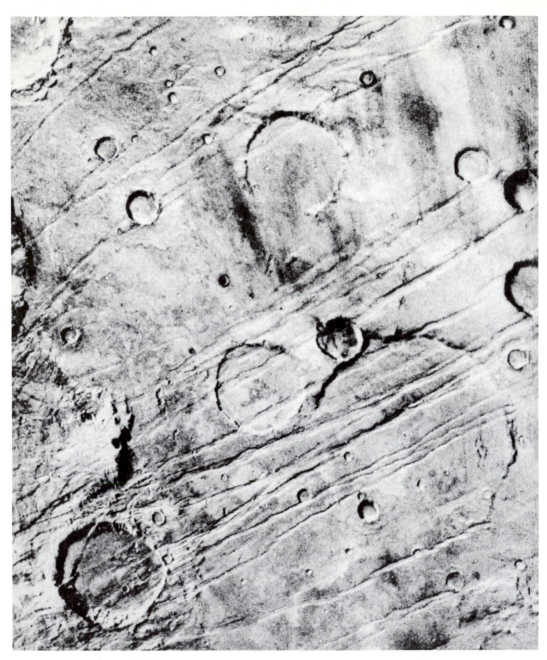

Figure 4.37
Viking Orbiter view of the southern cratered hemisphere of Mars (45°S, 95°W) showing extensive grabens that cut the plains and older craters; note, however, that many of the smaller, younger craters are superimposed on the grabens. Width of area shown is about 240 kilometers.

Regional Subsidence and Uplift

Grabens are indications of local extension of the crusts of a planet. Such extension generally results from either regional subsidence or uplift. The orientation of the lunar rilles that surround circular maria is indicative of regional extension of each circular mare and associated basin. The well-developed positive gravity anomalies—mascons—are also associated with many of the circular maria. One explanation for these relationships is that the weight of the emplaced lava caused subsidence in circular basins, expressed as the development of the grabens in the older lava deposits. But by the time the youngest lavas formed, the crust had cooled and thickened sufficiently to support the excess mass without further deformation. Thus both the development of grabens and the retention of a mascon can be ascribed to the response of the lunar crust to the emplacement of mare plains, rather than to the earlier basin-forming impact. On the other hand, the origin of the crustal extension that produced highland grabens remains unclear.

The plains that fill to overflowing the 1300-kilometer Caloris Basin on Mercury exhibit ridges and fractures oriented both radially and concentrically with respect to the center of the basin (Figs. 6.5, 6.8). The fractures appear to have formed after the ridges. These relationships are unlike anything elsewhere on the half of Mercury that has been explored so far or on the Moon. Subsidence of a circular feature as large as Caloris (whose diameter is about one-twelfth the circumference of the planet) will lead to compression rather than extension, because planetary curvature is significant. If such subsidence is postulated to explain the development of the ridges (through compression), then subsequent uplift—doming—is implied by the pattern of grabenlike features. Whatever the explanation, a crustal response quite different from that of the basins and related mare of the Moon is indicated.

Subsidence can be caused by withdrawal of materials other than lava, as is suggested by the chaotic terrains on Mars (Fig. 4.38). There, a large area of what had been ancient, heavily cratered terrain was later broken up, apparently as the result of withdrawal of some underlying substance; eolian and perhaps other kinds of erosion have subsequently removed parts of the surface, leaving a variety of bizarre landforms. The nature of the material initially withdrawn is uncertain. One possibility is that a large amount of subsurface ground-ice was volatilized; another is that volatile material such as carbon dioxide was emplaced in the soil early in martian history and subsequently was lost because of exposure to the current atmospheric conditions. Magmatic withdrawal also is conceivable. Whatever happened below the surface, subsidence very likely is the cause of the fracturing of the chaotic terrains.

The gigantic Valles Marineris canyon of Mars (Fig. 3.19) represents an enormous deficiency of material in the equatorial regions. The general east-west orientation of Valles Marineris appears to reflect large-scale extensional faulting. It is unclear whether the original material could have been reduced to small grains and then transported from the equatorial regions to the polar regions by wind or whether there has been substantial subsidence due to the withdrawal of magma at great depth beneath the entire Valles Marineris area. The proximity of the Valles Marineris depressions to the very large volcanoes of the Tharsis region to the west has been cited as a possible clue to regional magma withdrawal.

The Bulges on Mars

The dominant structural feature on Mars is an enormous bulge with a diameter one-third that of the planet's circumference, the Syria Rise (Fig. 4.14). This feature is centered near the equator, where three of the four huge martian shield volcanoes form the Tharsis Ridge. The rise is about 5000 kilometers in diameter and stands 7 kilometers above the average planetary surface. Farther west the somewhat older Elysium volcanoes cap a smaller bulge 2000 kilometers in diameter and 2 to 3 kilometers

Figure 4.38
Chaotic terrain on Mars stands out strikingly in the midst of cratered terrain in the Simud Vallis region in this Viking Orbiter mosaic. Arcuate fractures are indicative of subsidence. The subsidence here probably is related to a ground-sapping process rather than withdrawal of magma. Area shown is about 1200 kilometers across.

high. Crustal extension on a grand scale accompanied the development of the Syria Rise. Between the Tharsis volcanoes and the western end of Vallis Marineris is a region of gigantic extensional fractures, Noctis Labyrinthus. This is the highest region in the area and can be easily interpreted as the apex of uplift and extension. Other examples of large-scale extensional faulting occur throughout the Syria Rise, arranged radially with respect to the upwarp.

On Earth, most mountain ranges and all the continents are composed of lower-density material than that of surrounding crustal regions. They can be regarded as "floating" in a denser medium, such that their excess height does not imply excess weight that must be supported rigidly by the lower crust and mantle. One consequence of this relationship—termed isostasy—is that the gravity field of the Earth is relatively "smoother" than the topography, because height variations are compensated by density variations, at least on a large scale. It is therefore surprising that the Syria Rise is the site of a large, positive gravity anomaly, much greater in areal extent than the lunar mascons and quite probably reflecting density differences at depth as well as near the surface. Thus either the lithosphere of Mars must be extraordinarily thick to be capable of supporting the excess mass of the Syria Rise without density adjustment or there is a general upwelling of dense material persisting deep within the planet beneath the Syria Rise. We prefer the latter interpretation (i.e., internal convection) because it is a plausible way to account as well for the enormous quantity of basalt and the extraordinarily enduring volcanism there.

Evidence of Compression

The mare plains of the Moon, the similar lava plains on Mars, and the probable volcanic plains on Mercury all display ridges. These are low-lying, commonly sinuous, positive features, which, at least on the Moon, may be partly a product of local compression associated with the emplacement and cooling of the lavas. As for the Caloris Basin of Mercury, the ridges as well as a younger set of fractures cutting them are clearly organized in radial and concentric patterns relative to the basin, again implying a regional pattern of deformation during and after filling of the basin. On Mars, younger eolian deposits obscure much of the detail of the volcanic plains. Although ridges exist, it is difficult to deduce regional patterns from the existing data. Whatever the origin, the ridges of the plains do appear to reflect local deformation of the plains units rather than crustal contraction on a global scale.

On Mercury, other linear features recognized in the older terrains resemble thrust faults on the Earth. In a few places in the Mariner 10 photographs of Mercury, there appears to be compressional distortion of craters as a result of deformation by scarplike linear features (Fig. 4.39) that are interpreted to be thrust faults. The crustal contraction required to produce these features probably could have been accomplished with a decrease in planetary radius of the order of a few tenths of a percent; that is, if these global scarps are manifestations of a contracting crust and if the displacement can be based upon the few known cases of visibly deformed craters. No evidence of such global contraction exists on Earth, although most of the present surface of Earth records the deformation of only the past half-billion years. Likewise, no similar pattern has been recognized on the Moon or Mars. The mercurian features may be associated with the small shrinking (billions of years ago) of the uniquely large core believed to constitute most of the interior of Mercury. Such shrinking conceivably would have caused surface compression. In addition, global stresses associated with the slowing of planetary spin (due to tidal torques of the Sun) may have influenced the global orientation of the thrust faults.

Other Global Patterns

Telescope observations of the Moon long ago identified a characteristic pattern of linear breaks, alignment of crater walls, ridges, and other features that were collectively termed the *lunar grid*. The polygonalization of the walls of older craters is

Figure 4.39
Discovery Scarp, seen in this four-frame Mariner 10 mosaic, crosses nearly 500 kilometers of intercrater plains on Mercury. The two large craters cut by the scarp are about 45 and 30 kilometers in diameter. Maximum relief across the scarp is about 1 kilometer.

a characteristic feature. Polygonalization also characterizes Arizona's Meteor Crater where it is associated with a pre-existing jointing pattern in the host rocks. Evidently the continued action of gravity on the crater walls over time modifies them by slump and slide to conform preferentially to pre-existing zones of weakness. Presumably a similar effect has been operating in the lunar crust to account for the grid, although most of the features mapped as part of the grid probably are associated with the formation of the large impact basins and therefore do not really qualify as part of a global system of features. Ancient tidal interaction with Earth was one early suggestion regarding the origin of the lunar grid.

The characteristic plate-tectonic pattern of Earth differs markedly from anything seen on the Moon, Mercury, or Mars. It is associated with special circumstances below the crust that permit rigid plates on the surface to rotate and slide around alongside each other. Their motion may reflect the action of convection cells within the mantle. A paramount question in planetology is whether Venus exhibits similar surface features. The only pertinent evidence at present is provided by low-resolution radar maps acquired by Earth-based radar and by the Pioneer Venus Orbiter (Fig. 4.40). The large circular features seen in the radar images have been interpreted speculatively either as ancient impact craters or as possible sites of massive volcanic action. In addition, some of the linear features seem to resemble rift zones and other manifestations of plate tectonics on Earth. The similarities and differences of major tectonic features on Venus with those of Earth eventually will provide an extraordinary opportunity to learn about the inner workings of both planets. Similarly, the question of whether the large circular structures are ancient impact scars or enormous volcanoes could be tested directly.

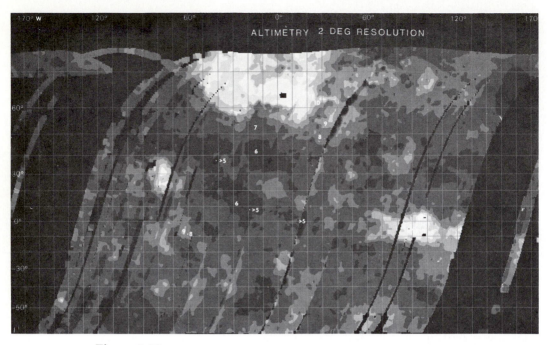

Figure 4.40
In this topographic map of Venus obtained by the Pioneer Venus radar altimeter, showing 83 percent of the surface of Venus, bright areas are high and dark areas low. Numbers indicate relative elevations in kilometers. Total relief approaches 15 kilometers.

It is the interplay of renewal from within and modification from without that produces the complex set of topographic and geological features that characterize the surfaces of the planets. The careful unraveling of those features and their origins is the challenge for the geologist in trying to understand the histories of these objects.

SUGGESTED READING

Anderson, D. L. "The San Andreas Fault." *Sci. Am.* **225**(5):52–68 (Nov. 1971). (Offprint No. 896.)

Bird, J. M., and B. Isacks. *Plate Tectonics: Selected Papers from the Journal of Geophysical Research.* Washington, D. C.: American Geophysical Union, 1972.

Burke, K. C., and J. T. Wilson. "Hot Spots on the Earth's Surface." *Sci. Am.* **235**(2):46–59 (Aug. 1976). (Offprint No. 920.)

Carr, M. H. "The Volcanoes of Mars." *Sci. Am.* **234**(1):32–43 (Jan. 1976.)

Carr, M. H., R. Greeley, K. R. Blasius, J. E. Guest, and J. B. Murray. "Some Martian Volcanic Features as Viewed from the Viking Orbiters." *J. Geophys. Res.* **82**:3985–4015, 1977.

Carr, M. H., and R. Greeley. *Volcanic Features of Hawaii—A Basis for Comparison with Mars.* NASA SP-403, 1980.

Cotton, C. A. *Volcanoes as Landscape Forms.* Christchurch, New Zealand: Whitecombe and Tombs, 1944.

Dewey, J. F. "Plate Tectonics." *Sci. Am.* **227**(5):56–68 (May 1972). (Offprint No. 900.)

Dietz, R. S. "Geosynclines, Mountains, and Continent-Building." *Sci. Am.* **227**(5):30–38 (Mar. 1972). (Offprint 899.)

Greeley, R., and P. D. Spudis. "Volcanism on Mars." *Rev. Geophys. Space Phys.* (in press).

Green, J., and N. M. Short. *Volcanic Landforms and Surface Features.* New York: Springer-Verlag, 1971.

Head, J. W. "Lunar Volcanism in Space and Time." *Rev. Geophys. Space Phys.* **14**:265–300, 1976.

Heirtzler, J. R., and W. B. Bryan. "The Floor of the Mid-Atlantic Rift." *Sci. Am.* **233**(2):78–90 (Aug. 1975). (Offprint No. 918.)

James, D. E. "The Evolution of the Andes." *Sci. Am.* **229**(2):60–69 (Aug. 1973). (Offprint No. 910.)

Macdonald, G. A. *Volcanoes.* Englewood Cliffs, N. J.: Prentice-Hall, 1972.

Malin, M. C. "Comparison of Volcanic Features of Elysium (Mars) and Tibesti (Earth)." *Bull. Geol. Soc. Am.* **88**:908–919, 1977.

Mutch, T. A., R. E. Arvidson, J. W. Head, III, K. L. Jones, and R. S. Saunders. *The Geology of Mars.* Princeton Univ. Press, 1976.

Phillips, R. J., and E. R. Ivins. "Geophysical Observations Pertaining to Solid State Convection in the Terrestrial Planets." *Phys. Earth Planet. Int.* **19**:107–148 (June 1979).

Phillips, R. J., and K. Lambeck. "Relationships Between Long-wavelength Gravity Anomalies and Tectonics on the Terrestrial Planets." *Rev. Geophys. Space Phys.* (in press).

Press, F., and R. Siever. *Earth* (2nd ed.). San Francisco: W. H. Freeman and Company, 1978.

Schultz, P. H. *Moon Morphology.* Austin: Univ. Texas Press, 1976.

Siever, R. "The Earth." *Sci. Am.* **233**(3):82–90 (Sept. 1975).

Tazieff, H. "The Afar Triangle." *Sci. Am.* **222**(2):32–40 (Feb. 1970). (Offprint No. 891.)

5

HISTORY OF
THE MOON

HISTORY OF THE MOON

The 1960s and early 1970s will be remembered for human-kind's first tentative steps into space. Born out of "cold war" competition, the American decision to send men to the Moon resulted in a marshaling of governmental resources on a scale unprecedented for a single peacetime objective. Scientists and engineers of nearly all disciplines were brought together to form a unique team that focused on the objective of a manned landing on the Moon by the end of the decade. Project Apollo, including the development of the Saturn V rocket, was—and remains—the largest single technological endeavor ever undertaken, surpassing even the World War II development of the atomic bomb.

Before undertaking manned flights to the Moon, it was necessary to gather as much information as possible on the nature of the surface of the Moon. Thus the Ranger, Surveyor, and Lunar Orbiter automated missions were developed and flown, supplemented by extensive Earth-based observations. Although the general priorities were to gather engineering data specifically for manned lunar landings, a tremendous wealth of scientific data were obtained as well. For example, of the five Lunar Orbiters—spacecraft designed to obtain high-resolution photographs of the lunar surface—the first three were so successful in photographing potential landing sites that the remaining two were reassigned mainly to map the nearside with contiguous pictures, to photograph sites of high geological interest, and to provide important new coverage of the farside of the Moon. The photography, obtained in 1966–1967, still provides the only regional coverage of the Moon. With the Apollo missions came thousands of additional pictures, in situ geophysical measurements, orbital geochemical and geophysical information, and an invaluable cargo—pieces of the Moon's own surface. What have we learned from this information, and how have our ideas of the Moon changed during the last decade? How have the lessons of the lunar program provided the framework for exploration of the other planets?

MYSTERY OF BIRTH

Traditional Views of Origin

Before the analysis of the lunar samples, it was not clear whether the Moon was chemically differentiated. Indeed, one principal view was that it accreted while cold and was never heated sufficiently to melt completely, thus remaining essentially homogeneous throughout. Some scientists held different views, a few even contending that the Moon had granite and other Earthlike crustal rocks, making it a miniature Earth except for the absence of a substantial iron core, which is ruled out by its much lower bulk density and high moment of inertia. The mechanism of formation of the unique Earth-Moon system was (and still is) highly speculative (Fig. 5.1). In principle, the Moon could be sibling, child, or spouse of Earth. That is, it could have originated by (1) separate accretion from the same planetary system, thus forming as a binary planet from the beginning; (2) fission from the proto-Earth; or (3) capture from an entirely different place in the Solar System.

The scientific rationale for the Apollo program—and especially the gathering of lunar samples for sophisticated chemical, mineralogical, and petrologic analysis—was organized from the beginning around an attempt to deal with these questions of chemical and dynamic origin.

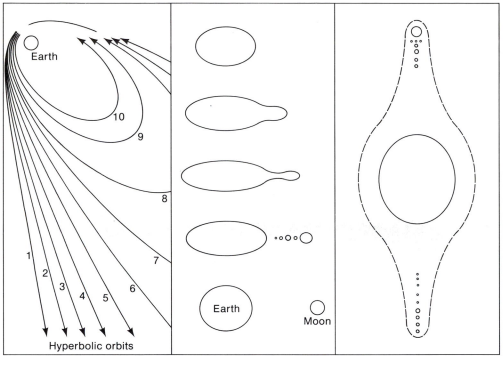

Capture	Fission	Binary accretion

Figure 5.1
Mechanisms for lunar formation. Schematic illustrations of the three theories of lunar origin. **(a)** Capture: selective capture by the Earth of bodies that accreted to form the Moon; fragments in orbits 1 to 5 are lost in hyperbolic orbits; those in orbits 6 to 10 are captured in elliptical orbits. **(b)** Fission of the Moon from "mother" Earth as it was spun to instability during core formation. **(c)** Binary accretion of both the Earth and Moon from the solar nebula. The Earth was formed first, and heavy metals settled to form a core. Lighter fragments in geocentric orbits accreted by collision to form the Moon.

The Testimony of Apollo

Perhaps the most unexpected result of the laboratory analysis of the samples is that the cratered topography of the highlands, once almost universally presumed to be the scars of lunar accretion dating from about 4.5 billion years ago, is instead the product of a later phase of global erosion, metamorphism, and impact reworking, caused by massive bombardment ending

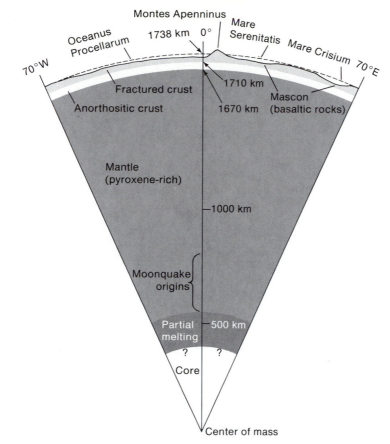

Figure 5.2
This cross section of the Moon shows a crust 65 kilometers thick, with a highly fractured upper 25 kilometers, a mantle 11 kilometers thick, and a core that is either partially molten or includes a molten zone. This cross section represents a segment of the Moon drawn to scale from a sphere 1738 kilometers in radius from the center of the mass. [After "The Moon after Apollo" by Farouk El-Baz, *Icarus*, vol. 25, fig. 17, p. 528, August 1975. Academic Press, Inc.]

over half a billion years after the Moon first formed as a solid object. Moreover, the lunar samples show that the Moon is chemically differentiated. The evidence is relatively strong that segregation into chemically distinct crust, mantle, and possibly even a small core (Fig. 5.2) must have occurred very early in lunar history. Thus a period of perhaps half a billion years of the Moon's early history remains obscure, bracketed by its early chemical differentiation and by the end of heavy bombardment.

The crust of the Moon is relatively enriched in potassium, uranium, and thorium—naturally radioactive elements that give off heat. If the Moon as a whole had the same composition, it would be entirely molten at present. Thus it is clear that the 60- to 100-kilometer-thick lunar crust is chemically distinct from the remainder of the planet. In addition, isotopic evidence demonstrates conclusively that the separation of the crust took place at least 4.4 billion years ago, nearly coincident (or actually coincident) with planetary birth.

The lunar crust is depleted by a factor of 10 to 100 in the elements that are chemically associated with iron (due to similarities in effective ionic size and charge) in the formation and evolution of a planet. These elements are referred to as *siderophilic,* as distinguished from *lithophilic*—those associated with the silicate minerals. The lower density of the Moon suggests that it is composed of lithophilic elements more or less separated from the siderophilic phase. In this respect, the Moon differs profoundly from the Earth. The evidence is strong that this depletion took place during or possibly even before the Moon accreted as a body, suggesting early chemical separation in the solar nebula.

Other distinct chemical differences exist between Moon and Earth. For example, the potassium/uranium ratio (Fig. 5.3)—a good index of classes of terrestrial rocks and meteorites—is consistently lower for lunar samples than for terrestrial samples. A similar relationship holds for the lead/uranium ratio. In addition, the Moon is highly deficient in certain elements like sodium (Fig. 5.4). Indeed, the chemistry of the Moon, especially with regard to the trace elements, indicates a systematic enhancement relative to Earth of *refractory* lithophilic elements—that is, elements that survive high-temperature volatilization and are retained to form silicate minerals in the resulting cooled planetary mass. These relationships demonstrate that the Moon and the Earth formed in two fundamentally different chemical "crucibles." If they shared a common history at some point, a subsequent high-temperature chemical volatilization process must have operated on the lunar material before it collected to form the actual Moon. Moreover, the lunar material is easily distinguished chemically from meteorites, which

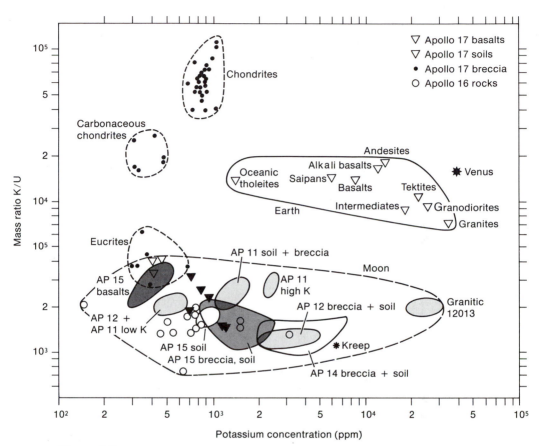

Figure 5.3

Comparison of potassium/uranium ratios and potassium concentration for lunar rocks, terrestrial igneous rocks, chondrites, and eucritic meteorites. All of these Solar System objects have remarkably well-defined K/U ratios. It appears that this ratio and other lithophile element ratios are approximately preserved in magmatic processes and can be regarded as "planetary constants." [After "Primordial Radioelements and Cosmogenic Radionuclides in Lunar Samples from Apollo 15" by G. D. O. Kelley et al., *Science*, vol. 175, fig. 2, January 28, 1972. Copyright 1972 by the American Association for the Advancement of Science.]

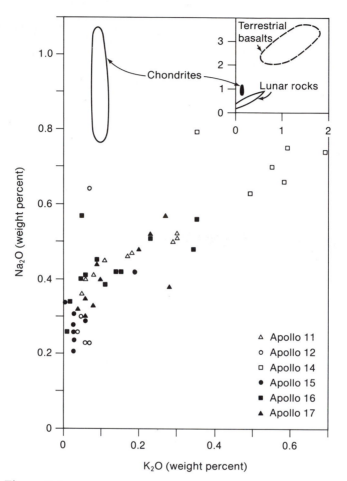

Figure 5.4
The field described by the various chondrite classes is indicated in this correlation diagram of Na_2O versus K_2O in lunar rocks (individual data points). The inset is the same plot on a larger scale; the area enclosed by the broken line indicates the average of various terrestrial basalt subtypes. The terrestrial field may overrepresent alkali basalt types; nevertheless, the alkali-poor character (particularly for Na) of many lunar rocks is shown. The region south of Mare Imbrium is rich in K, as shown by the Apollo 14 rocks. [From "Lunar Science: The Apollo Legacy" by D. S. Burnett, *Reviews of Geophysics and Space Physics*, vol. 13, fig. 16, July 1975. Copyrighted by the American Geophysical Union.]

Plate 1
Surface of Mars viewed from Viking Lander 2. Mars, the red planet, gains its color from a nearly ubiquitous blanket of fine-grained red dust. This view of Utopia Planitia shows a surface littered with blocks of rock, a salmon-colored sky caused by dust particles suspended in the atmosphere, and various features of the Viking spacecraft.

Plate 2
Ice on Mars. This high-resolution color photograph of the surface of Mars, taken by Viking Lander 2, shows a coating of ice on the rocks and soil. The time at which this frost appeared corresponds almost exactly with the buildup of frost one martian year (23 Earth months) previously. This coating of ice, like that which formed one martian year ago, is extremely thin, probably less than a few microns in thickness. Its composition is suspected to be water ice.

Plate 3

Jupiter and its four planet-size moons, the Galilean satellites. Separate pictures were photographed in early March of 1980 by Voyager 1 and assembled into this collage. They are not to scale but are in their relative positions. Reddish Io (upper left) is nearest Jupiter; then come Europa (center), Ganymede, and Callisto (lower right).

Plate 4
Io viewed by Voyager. Io, about the same size and mass as Earth's Moon, is yellow orange, partly because of the high sulfur content of its surface. The circular features seen here are mostly volcanic calderas.

Plate 5

Volcanic explosion on Io. An enormous volcanic plume can be seen silhouetted against dark space as much as 100 kilometers above Io's bright limb. The brightness of the plume has been increased by computer enhancement, but the relative color of the plume (greenish white) has been preserved. Volcanic explosions similar to this occur on Earth when magmatic gases are vented. On Earth, water is the major gas driving such explosions. Because Io is thought to be extremely dry, other gases are presumably involved.

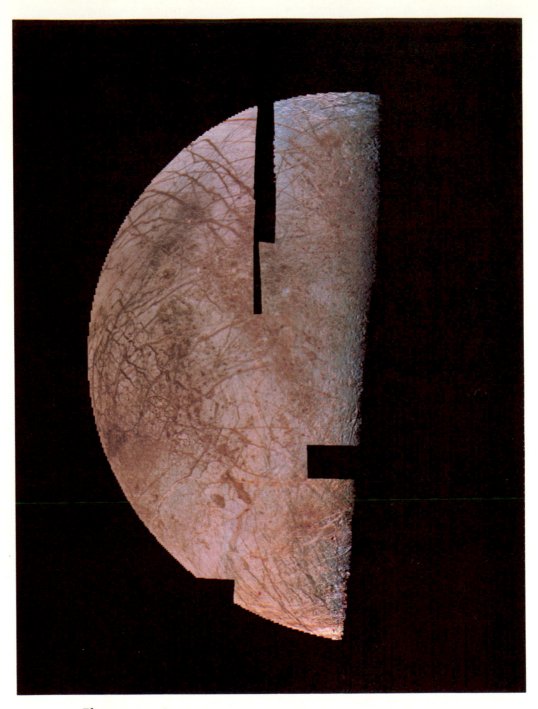

Plate 6
Europa viewed by Voyager 2. This computer-generated mosaic of Europa shows dark, rough, mottled terrain; bright, smooth terrain; and cycloid ridges. The extremely low frequency of impact craters suggests that the surface of Europa is relatively younger than those of Callisto and Ganymede.

Plate 7

Color mosaic of Ganymede. The density of this Mercury-sized satellite of Jupiter is less than 2 grams per cubic centimeter, suggesting a composition that includes substantial amounts of water. Two distinct terrains are visible here: dark, heavily cratered, ancient terrain is cut by networks of lighter, "grooved," younger terrain.

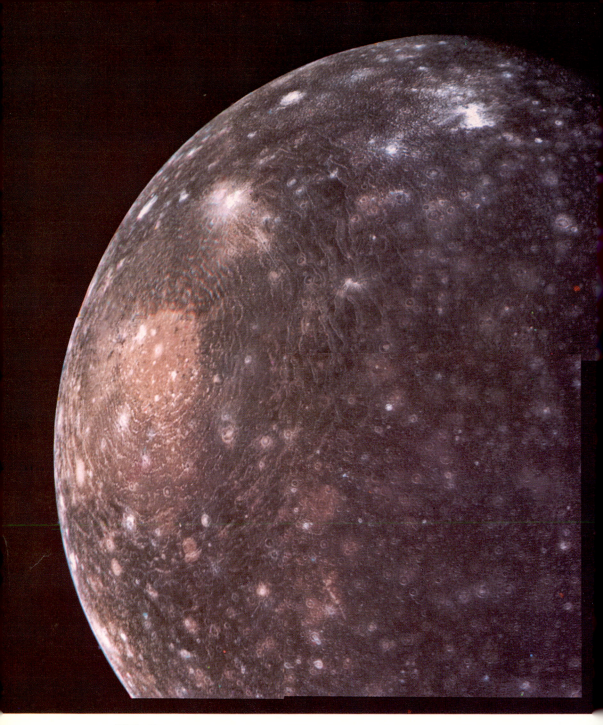

Plate 8
Callisto viewed by Voyager 1. Callisto, also about the size of Mercury, is composed of a mixture of water and silicates. Of the Galilean satellites, Callisto is the most heavily cratered.

suggests that the products of perhaps many different, unique chemical crucibles throughout the Solar System await discovery.

A final enigma posed by the lunar samples and the related orbiting satellite data is that although the Moon as a whole exhibits a very low magnetic field—less than a ten-thousandth that of the Earth—certain areas on the Moon exhibit as much as 1 percent or more of the Earth's field. No plausible explanation has been advanced for these comparatively strong *remanent* magnetizations. They are generally regarded as fossil magnetizations surviving somehow from the time of the Moon's formation. Perhaps there was an internal dynamo similar to the Earth's and the present field was "frozen in" as the Moon cooled. Or perhaps the Earth's field was stronger and influenced the early lunar crust—at a time when the two bodies were nearer to each other.

In the search for the origin of the Moon, not only were chemical questions answered (and new insight gained into the Earth's formation), but a totally unanticipated and otherwise undetectable phase of Solar System history was also discovered: the late heavy bombardment of the Moon, dated at 4 billion years—a phase that undoubtedly affected the Earth as well. Yet, to unravel the full story of the birth of the Moon, chemical analyses of samples from other planets will be required, so that the special effects associated with the Earth-Moon system can be distinguished from processes that operate throughout the entire Solar system.

Planetary Stratigraphy

Even without chemical samples, some clues to the makeup of planetary surfaces and geological history can be derived from orbital photography alone. The primary observational data in a photograph are landforms—craters, ridges, scarps, mountains, plains. The first task is to recognize and classify them and, where possible, to infer the processes responsible for their origin. The result of placing such kinds of information on a geographic base map is a *terrain map*, a physiographic representation of the planetary surface.

Terrain maps, however, show only *surface configurations*. Geological mapping extends the analysis beneath the surface to show the distribution of three-dimensional *rock units* and to place them in a relative sequence. For example, the units that form the lunar maria can be recognized on photographs as continuous; they overlie other units, and in some cases are overlain by yet younger units. Hence, even before Apollo, it was possible to map some rock units and to construct a relative sequence of ages. Geological maps, when combined with interpretations of the processes that formed the mapped rock units, lead to the derivation of a geological history.

In the remainder of this chapter, this approach is illustrated, supplemented, and supported by the absolute dates obtained from Apollo lunar samples. In later chapters, we discuss the relative sequences of events for Mars and Mercury, derived primarily from photographs, with the Moon serving as the "yardstick" for intercomparisons.

BASINS UPON BASINS—THE STORY OF THE HIGHLANDS

From Earth, observation of the Moon easily reveals two distinctive physiographic provinces—the dark maria and the bright highlands (Fig. 2.1). Early geological mapping of the Moon (through telescopes) established the relatively old age of the highlands and the fact that they are dominated by large impact basins.

Studies of early lunar history revolve about two interrelated problems: (1) the absolute time scales for the formation of materials and landforms and (2) the rate of impact bombardment.

Lunar Chronology—How Old Are the Rocks?

The most common type of rock fragment sampled in the Apollo program is itself composed of rock fragments and is termed a *breccia*. The three principal types of samples collected on the Moon are (1) crystalline fragments of old rocks, such as basalts, which formed within the crust, (2) breccias formed by the impact of meteorites, and (3) soils. Highland samples are dominated by breccias; in fact, some highland rocks are breccias that are themselves incorporated into breccias (Fig. 5.5). It is be-

Figure 5.5
The thin section (lower picture) of an Apollo 16 sample (upper picture) showing multiple fragments of breccias within a brecciated sample, attesting to the complex history of impact-associated rock formation.

lieved that objects impacting the Moon during its early history fractured and fragmented the surface, producing a layer of fragmental debris that became compacted and metamorphosed into rock by the heat and pressure generated by the passage of shock waves associated with continued bombardment.

When did this bombardment occur? Let us begin by addressing the more general subject of how we date lunar rocks. Four methods are currently used to determine the ages of lunar samples. All rely on the process whereby unstable radioactive isotopes decay at known constant rates to form stable, lighter isotopes. One method measures the amount of stable lead produced by the decay of radioactive uranium and thorium. A second measures the amount of stable strontium produced by the decay of radioactive rubidium. A third measures the amount of argon gas produced by the decay of radioactive potassium. A fourth measures the amount of stable neodymium produced by the decay of radioactive samarium.

To determine the age of a sample, the amounts of radioactive parent isotope and stable daughter isotope are measured. Combined with measurements of decay rates, these data allow the length of decay time to be estimated, which would be the "age" of the material since it crystallized—if there were no other complications. However, the unique interpretation of this "age" is complicated by: (1) limits to our knowledge of the initial abundances of parent and daughter elements; (2) errors in the decay-rate values; and (3) the possibility that heat, pressure, and shock during the geologic past may have affected the daughter element differently than the parent one.

Lunar Cratering—When Did It Occur?

It has been known for centuries that the Moon displays two distinct landscapes—one heavily cratered and the other sparsely cratered. Most of the large craters are the result of impact by asteroids and comets. Thus the number of superimposed craters (per unit area) should be proportional to the age of the surface and should be an index of relative ages for different areas of the Moon.

The actual ages determined by isotopic analysis of samples representative of several units on the Moon can be used to establish the absolute time scale of lunar bombardment chronology (Fig. 5.6). The narrow range in ages for mare rocks can be

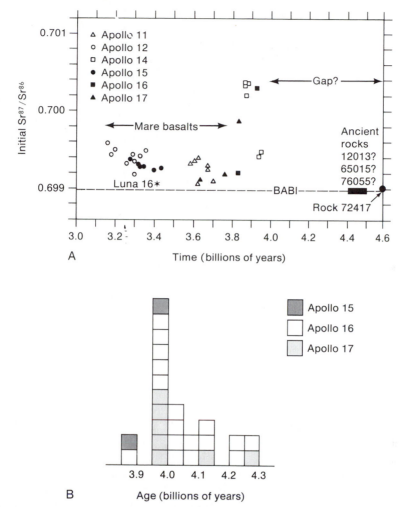

Figure 5.6

In the upper figure, ages of lunar samples are plotted as determined by measurement of the amount of Strontium 87 produced by the decay of radioactive Rubidium 87. Note that the mare basalts fall between 3.15 and 3.80 billion years old and that, with the exception of a few samples, all materials returned by the Apollo and Luna missions were younger than 4.0 billion years old. It is on the basis of these data that a "terminal cataclysm" was proposed, suggesting that all the craters and large basins currently visible on the lunar surface formed during a brief "event" about 4 billion years ago. However, in the lower figure, potassium-argon isotopic age determinations (based on the amount of Argon 40 produced from radioactive potassium 40) seem to show a greater range in ages, including ages between 4.0 and 4.3 billion years. More refined analyses of uranium-lead, neodymium-samarium, rubidium-strontium, and potassium-argon ages, as well as theoretical models, now seem to suggest that cratering occurred throughout the period between 4.0 billion years ago and the time of formation of the lunar crust, about 4.4 billion years ago.

interpreted as the age of lava flows in the areas sampled. Of much greater controversy is the meaning of the peak in highland ages at 4 billion years and the apparent underabundance of certain ages greater than 4 billion years seen in the rubidium-strontium data (Fig. 5.6). Is this a sampling problem (due perhaps to chance positions of a few dominant impact basins) or does it represent a part of lunar history that is "lost"?

The Early Bombardment History of the Moon

Various interpretations of the apparent radiometric ages of highland breccias have led to three main theories concerning lunar bombardment. One theory proposes that the observed impact rate represents only the tail end of a continuous history of declining bombardment resulting from the accretion of the Moon (continuous bombardment). A second theory, based on the concentration of ages at 4 billion years, suggests a terminal cataclysm of bombardment in which most of the largest basins were formed during a relatively short period of time (cataclysmic bombardment). The third hypothesis suggests that the 4-billion-year age reflects a reduction in the formation of large craters, not basins, and that basin bombardment might have been either continuous or cataclysmic, but that the age data do not reflect that bombardment (large crater bombardment). A controversy has developed between the proponents of the continuous and cataclysmic lunar bombardment schemes. The continuous bombardment interpretation was proposed first. Soon after the first three lunar Apollo missions, however, analyses of large numbers of highland samples from regions widely distributed over the Moon all turned out to show a peak age at 4 billion years. Moreover, there was an apparent lack of samples older than this age. Thus a "terminal cataclysm" was proposed, and the argument developed as to whether the cataclysm consisted of the formation of many or all of the large basins or just of the last and largest basins—Imbrium and Orientale. If no sample older than 4 billion years could be found at any of the Apollo or Luna sites, this would lend credibility to the former position. If some samples of different ages could be found, the latter position would gain support. It is here that the interpretations of radiometric ages differ. The rubidium-strontium method of dating shows very few samples with ages greater than 4 billion years and is considered to be the strongest evidence for cataclysmic bombardment. However, the uranium-thorium-lead and potassium-argon methods do show evidence of older processes. What is probably involved are poorly understood differential effects of heating due to impact and other processes. As a consequence, the data cannot be judged to favor any one hypothesis decisively over the others.

What Else Happened to the Moon?

Although bombardment clearly played a dominant role in shaping the ancient surface of the Moon, other things were happening at the same time. Chemical evidence indicates that the Moon melted, differentiated, and formed a crust very early in its history, including an unusual low-density anorthositic rock (composed almost entirely of a calcium-rich plagioclase mineral), which may have "floated" on a mantle of denser magnesium- and iron-rich silicates. Partial melting of the lower crust produced several unusual materials, the most important being a differentiated rock rich in potassium (K), rare earth elements (REE), phosphorus (P), barium (Ba), uranium (U), and thorium (Th). Called KREEP, this material has enriched some breccias and soils with an "exotic" component of high radioactivity. KREEP materials may be the product of a very early form of volcanic activity on the Moon, and thus are important to the understanding of early thermal processes.

Controversy remains the key word in studies of the early lunar history. After the planet differentiated, it experienced heavy bombardment over a period of nearly 500 million years. Regardless of whether this bombardment was continuous or episodic, volcanism occurred as well throughout much of early lunar history.

THE IMBRIUM AND ORIENTALE BASINS

The Imbrium and Orientale basins have been studied more extensively than any of the other 40 basins on the Moon. Both
were formed late in the period of heavy bombardment and display relations that are important to the understanding of basin
geology.

Imbrium—Window into the History of the Moon

Prominently positioned in the nearside northern hemisphere
(Fig. 2.1), the Imbrium Basin is the key to the geology of more
than a fifth of the lunar surface. Much of the early photogeological lunar mapping was done on parts of the Imbrium
Basin because its stratigraphic and structural relations are well
displayed and allow relatively unambiguous mapping. In fact,
most of the photogeological techniques later used for mapping
Mars' and Mercury's surfaces are derived from these early efforts.

The origin of the Imbrium Basin has been a subject of some
controversy. Gilbert first identified the distinctive radial
grooves and furrowing around the Imbrium Basin, which he
called *Imbrium sculpture* (Fig. 5.7), and correctly deduced an
impact origin of the basin. The main rim of the enormous crater
making up the Imbrium Basin (Fig. 2.1) is defined by some of
the most prominent mountain ranges on the Moon—the Alpes,
the Apennines, and the Carpathians—which in some places
rise more than 5 kilometers above the mare surface. These
mountain ranges outline a diameter of more than 1300 kilometers and in places are made up of rugged rectilinear massifs 10
to 30 kilometers across. They are interpreted to be parts of the
pre-basin crust that were uplifted at the time of basin formation
and then subjected to post-basin tectonic settling and deformation. Close examination of the Imbrium Basin reveals the remnants of two additional rings within the outermost chain of

Figure 5.7
In this oblique view looking north at a region in the central highlands, prominent lineated terrain is oriented radially to the Imbrium Basin; this characteristic surface morphology, termed "Imbrium sculpture," is thought to be related to low-angle secondary impacts and/or structural fractures associated with the formation of the Imbrium Basin. The large crater at right center is Herschel, approximately 40 kilometers in diameter.

mountains. The inner rings can be identified by the individual mountain blocks that rise above the mare surface and the circular arrangement of mare ridges within Mare Imbrium. The innermost ring is about 600 kilometers in diameter and is barely discernible.

The deposits associated with the Imbrium Basin can be separated into several different units. Closest to the basin is the Alpes Formation, a blocky or knobby, but smooth-surfaced terrain made up of randomly spaced hills 2 to 5 kilometers across. This material may consist of degraded ejecta, structurally deformed pre-basin rock, or a combination of both. The Alpes Formation grades radially outward into the Fra Mauro Formation, the main ejecta unit of the Imbrium Basin. This unit consists of sinuous-to-straight, smooth-textured ridges and elongate hummocks that are typically 2 to 4 kilometers wide and 5 to 20 kilometers long.

During discussions of possible landing sites for the Apollo missions, it became apparent that a landing on the Fra Mauro Formation should come early in the Apollo sequence because of its importance to understanding the geology of the Imbrium Basin. With sampling of this extensive unit as the prime objective, Apollo 13 was launched in 1970, but the explosion of an onboard oxygen tank nearly destroyed the spacecraft. The mission was then limited to the task of returning the astronauts safely to Earth. Subsequently, Apollo 14 was sent to obtain samples of the Fra Mauro Formation. The landing site was selected in a broad, shallow valley between radial ridges of the Fra Mauro Formation, north of the crater Fra Mauro and about 500 kilometers from the edge of the Imbrium Basin (Fig. 5.8). The primary sampling objective was Cone Crater, a small, fresh, sharp-rimmed impact crater 340 meters in diameter that penetrates deep into the formation. The plan was to make a sampling traverse of the Cone Crater ejecta field all the way up to the rim. The ejecta field of impact craters contains fragments of the surface rock layers; furthermore, the fragments tend to be distributed radially in terms of their original depth below the surface. Thus, by sampling from the outside of the ejecta field toward the rim, samples representing progressively deeper units of the formation could be obtained—a sort of poor man's core sample.

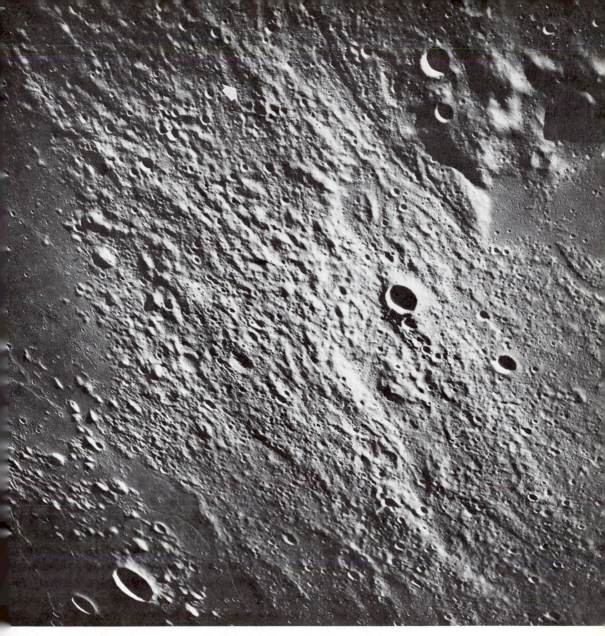

Figure 5.8
This photograph of the type area for the Fra Mauro Formation shows the continuous deposit ejecta from the Imbrium impact. Note the hummocky, irregular nature of the topography; the faint linear texture extending from the upper left to lower right is approximately radial to the Imbrium Basin. The Apollo 14 landing site is indicated by the arrow. The prominent crater near the center of the picture is Fra Mauro D, approximately 5 kilometers in diameter.

In their hike to Cone Crater through rugged terrain and over dusty surfaces, the astronauts found the traverse to be more strenuous than expected and were unable to reach the crater rim. However, they came within about 20 meters of it. Thus most of the sampling objectives were fulfilled. Of the approximately 43 kilograms of rocks and soil that were returned from the Apollo 14 mission, nearly all are breccias. Mixed with them are glasses and annealed rock fragments, plus some crystalline rocks. Interpretation and reinterpretation of the origins of these surface samples continue even today. It is likely that the samples include Imbrium ejecta, probably even some representatives of the uppermost crust in the Imbrium Basin area.

Orientale—The Best-Exposed Multiringed Basin

Unlike the Imbrium Basin, the Orientale Basin (Fig. 3.10) is not flooded with extensive mare basalt units and thus provides the opportunity to study the interior details of a multiringed basin. Orientale appears to be one of the youngest multiringed basins on the Moon and is available for study from orbital photography. Study of its structure allows interpretations to be applied to older and more degraded lunar basins and to multiringed basins on other planets.

A distinctive unit, the Hevelius Formation, has been mapped around the Orientale Basin and is interpreted as basin ejecta. The Hevelius Formation, like its analog the Fra Mauro Formation, consists of closly spaced ridges and troughs about 7 kilometers long and 1 kilometer wide, aligned radially or subradially to the basin. This part of the Hevelius Formation grades outward to a smoother terrain that subdues the underlying topography. In some places, the Hevelius Formation appears to have been emplaced as a dense blanket of debris overriding the rims of large craters (Fig. 5.9). At still greater

Figure 5.9
Debris overriding craters is indicated about 1100 kilometers southeast of the center of the Orientale Basin, where coarsely festooned ejecta deposits from the basin evidently moved from the upper left toward the lower right. Illumination is from the right; width of area shown is about 13 kilometers.

HISTORY OF THE MOON

distances, the ejecta deposits and secondary craters from the Orientale impact overlie the Fra Mauro Formation, indicating that the Orientale event took place after the formation of the Imbrium Basin.

The concentric mountain ranges (Fig. 3.10) of the Orientale Basin are of great interest to lunar geologists, and pose several interesting problems that are as yet unresolved. The outermost ring is made up by the Cordillera Mountains, which have a diameter of about 900 kilometers. Just inside this ring are the outer Rook Mountains, which are about 620 kilometers across and encircle a third, irregular mountain chain. Which of these three mountain rings marks the original rim of the crater? If the outer Rook Mountains constitute the crater rim, the original crater was about 620 kilometers in diameter. In this case, the Cordillera ring would be interpreted as a fault scarp that formed in the terminal stages of the cratering event, when the crust collapsed inward toward the transient crater, producing a "mega-terrace." On the other hand, much smaller diameters for the initial Orientale crater have been proposed, along with a variety of mechanisms to explain the formation of the multirings. For example, impact into the lunar crust may have generated tsunami-type waves that were frozen in place to form the concentric mountain rings. Regardless of the primary mode of formation, the asymmetric profile of the mountain rings shows that they have been modified by post-impact slumping and basinward downfaulting to produce prominent scarps. Similar basin-facing scarps are observed in preserved sections of the mountain chains around Imbrium and other basins.

Geologic mapping within the Cordillera ring reveals many diverse units. Between the Cordillera Mountains and Rook Mountains is a knobby, hummocky unit named the Montes Rook Formation. It is gradational with the Hevelius Formation in some places and can be interpreted as part of the ejecta sequence, which would support the interpretation that the Rook Mountains define the original crater boundary. The Maunder Formation occurs within the Rook Mountains and is characterized by hummocks and cracks. This unit is interpreted as impact melt that fell back into the basin, draping over pre-existing topography; cooling and subsequent drainage of the melt to lower areas caused the surface crust to crack and deform. Toward the basin, the Maunder Formation grades into a smooth-textured, light plains unit. Both the Maunder Formation and the light plains unit are cut by numerous fractures and grabens. Mare basalt units in Orientale are of limited extent and are superimposed on the older basin formations.

Gravity data obtained from the Apollo and earlier missions show that a mascon occupies the innermost ring of the basin, but the postulated volume of the mare fill is small compared to that of mascon basins closer to the sub-Earth point. Because there is little evidence of long-term, post-impact modification of the crater or extensive mare fill, it has been suggested that the Orientale mascon formed as part of the cratering event, not during the subsequent basaltic extrusion, as has been suggested for the filled circular basins elsewhere on the nearside.

Highland Plains

The highlands constitute more than 80 percent of the surface of the Moon. Early unmanned missions and many of the Apollo missions concentrated on the mare plains of the Moon, partly for engineering reasons, but also because mare areas have tended to attract more scientific attention in lunar studies than have the highlands. Although at one time the highlands were considered to be essentially uniform in character, we know now from orbital data and from analysis of lunar samples that the highlands are heterogeneous in terms of both morphology and geochemistry.

Three Apollo missions were planned to obtain information specifically on the highlands and old crustal rocks. The Apollo 15 mission to the Hadley region obtained samples of deep crustal material exposed in the uplifted blocks of the Apennine Mountains, part of the outermost ring around the Imbrium Basin (Fig. 5.10). The Apollo 17 site was selected in a valley within a highland region near the Serenitatis Basin to sample not only old crustal material but also what erroneously were considered to be young volcanic fillings. But the best example of a highland site was the Apollo 16 mission in the Descartes Highlands.

The Apollo 16 site was an area about 60 kilometers north of the crater Descartes. Lying west of the Kant Plateau, the landing area was selected to sample both the extensive Cayley Formation (Fig. 5.11) and the more rugged Descartes Mountain unit. The Cayley Formation is a widespread unit that has been

Figure 5.10
This view shows **(1)** 120-kilometers-long Hadley Rille, a sinuous rille that is interpreted to be a lava channel that was involved in the emplacement of mare lavas. Part of the Apennine Mountains **(2)** are seen in the lower center of the picture. These blocks form the outer rim of the Imbrium Basin and are parts of the lunar crust that were uplifted during the formation of the basin. Tectonic grabens in the form of linear rilles **(3)** are seen to both post-date and pre-date some of the mare basalts.

mapped in many areas of the lunar highlands. Characterized by relatively smooth surfaces of light tone, the differences in surface morphology from place to place suggest that it may consist of more than one unit and may have different origins. Many of the lunar scientists who participated in the discussions of the Apollo 16 site selection leaned toward the volcanic interpretation, and most geologic maps available before the Apollo 16 mission interpreted this unit as of probable volcanic origin.

Figure 5.11
The Apollo 16 landing site (star) was selected to sample the Cayley Formation, the smooth plains unit filling highland areas of the central highlands. Returned samples showed the materials to be of impact origin, not volcanic, as most investigators had thought prior to the mission.

Then came the samples from Apollo 16! Most of the samples proved to be impact breccias, ruling out a volcanic origin for the Cayley Formation, at least for the area sampled.

The two current hypotheses for the origin of the Cayley Formation are that it consists of: (1) primary ejecta excavated from the Orientale Basin as fluidized, ground-hugging debris that filled low-lying areas to form a smooth Moon-wide deposit or (2) a mixture of secondary crater ejecta and local material formed when Orientale impact rained down on the lunar surface. The difference between these two ideas is that in the first, most of the material was derived from a single, far-distant source, whereas in the second, most of the material was derived locally.

Evidence for ejecta-derived deposits is growing. For example, geochemical data remotely sensed from orbit show that the Cayley Formation is not only different from place to place, but that the Apollo 16 landing site differs chemically from deposits of both the Orientale and Imbrium basins. These facts argue against a widespread blanket of debris derived from the Orientale Basin and suggest instead that the unit was derived in place from local material. Moreover, calculations of the quantity of debris generated locally by secondary impact show that most of the material could be of local origin.

Another minor type of highlands unit is characterized by a blocky, jumbled surface morphology; two prominent patches occur on the lunar surface, one on the opposite side of the planet from the Imbrium Basin and the other on the opposite side from the Orientale. It has been speculated that ejecta from the formation of these basins created the two areas of the hummocky terrain. Studies of trajectories, however, suggest that relatively little material thrown from a major impact would travel halfway around the Moon.

An alternative interpretation is that this terrain was created by seismic energy generated by basin-forming impacts. Such events would generate moonquakes that would travel through and around the planet to cause large vertical displacements of the surface on the opposite side. For an event the size of Imbrium, the vertical displacement could have been at least 10 meters. Whatever the details, it is likely that impact-generated seismic disturbances play an important role in the degradation of planetary surfaces.

BASALTIC PLAINS—THE FORMATION OF THE MARIA

Maria cover about 17 percent of the lunar surface (Fig. 1.3). These flat, dark areas suggested to early observers that they might be bodies of water, hence the term *maria*—Latin, meaning seas. With improvements in telescopes, it became apparent that the maria were actually smooth, dark land surfaces. Speculation on their nature continued until the first samples were brought to Earth by Apollo 11 astronauts in 1969, although there were clues to the true character of the maria even before then. For example, Earth-based visual telescopic observations and photographs showed features that appeared to be lava flow lobes—an interpretation strengthened by high-resolution photographs taken by the Lunar Orbiter missions in 1967. In addition, the automated Surveyor Lander performed simple surface chemical analyses in 1966 that suggested that the composition of the maria most closely matched that of basalt, a common volcanic rock on Earth. Of the nearly half a ton of samples collected during the Apollo and Luna missions, most of the crystalline rock samples from the mare sites are in fact composed of dark-colored, fine-grained basalt, confirming the original interpretation of the Surveyor data.

Composition and Origin of the Mare Basalts

At one time, all maria were considered to be essentially the same. Several lines of evidence, however, show that this is not the case. First, crater abundances vary from mare to mare, indicating differences in times of formation; this has been confirmed by isotopic dating of mare samples, which range in age from about 3.2 to 3.8 billion years. Second, detailed Earth-based and orbital color-mapping delineates distinctive units, indicative of differences in composition and surface texture. Third, orbital geochemical mapping of gamma ray emission from the surface shows heterogeneities in composition of mare units. Fourth, rock samples from various mare sites show chemical, mineralogical, and petrographic differences. For example, the Apollo 11 and 17 basalts have titanium dioxide

(TiO_2) contents greater than 10 percent, whereas basalts from other Apollo mare sites range from 1 to 4 percent TiO_2.

Mare basalt samples range in texture from glassy (indicative of rapid cooling) through fine-grained vesicular to coarse-grained vesicular and nonvesicular gabbroic rocks (suggestive of long crystallization period at depth). Like terrestrial basalts, the typical mare basalt is composed of the minerals pyroxene, olivine, and plagioclase (with minor ilmenite, chromite, and other accessories). Compared to highland rocks, they are high in FeO and low in Al_2O_3.

Unlike terrestrial basalts, the lunar basalts lack chemically bound water and are low or totally lacking in volatiles and the major elements potassium and sodium. A basic, still-unanswered question is: Were the volatiles never accumulated on the Moon in abundance or were they distilled out at an early phase in lunar history?

Laboratory experiments with mare basalt samples show that they are most likely derived from an olivine–pyroxene-rich source rock by partial melting deep within the lunar mantle. The high-titanium basalts from the Apollo 11 and 17 samples appear to have been derived from a pressure-temperature regime around 100–150 kilometers deep, whereas the low-titanium basalts seemingly were derived from sources 200–500 kilometers deep.

Isotopic ages for mare basalts show that the earliest dated basaltic lavas were erupted about 3.83 billion years ago. (Even earlier mare volcanism is suggested by indirect evidence.) Sporadic eruptions continued until approximately 3.16 billion years ago. (Even younger basalts are suggested by low crater abundances in some unsampled units, where rocks may be as "young" as 2 billion years.) Thus the Moon was volcanically active in the mare regions for at least 700 million years—a time span greater than the entire period on Earth from the origin of the first organisms with "hard" parts to the development of man! Considering the relatively small total volume of mare lavas on the Moon, when averaged over the total time of eruption, the average rate of production is very low by terrestrial comparison. Photogeologic evidence, however, shows that the lavas were produced by both flood-type and plains-type eruptions. Since eruptions of both types have fairly high rates of effusion, the total period of lunar basaltic volcanism must have been characterized by lengthy periods of quiescence.

Topographic profiles derived from Apollo orbital experiments across highland and mare terrains suggest that mare basalts tend to occur in low-lying areas of the Moon (Fig. 5.12). Moreover, the relative elevations of the basalts among the individual basins seem to be similar, suggesting global hydrostatic control of the degree of flooding in each of the basins. When the figure of the Moon is compared to its center of mass, it is apparent that the center of mass is offset toward Earth. The mare basalts appear in this context perhaps to have tended to rise toward an equal elevation (gravitational) at all places on the Moon, Thus there may be a physical explanation for the fact that the maria are restricted to the lunar nearside. On the farside, where the "crust" is thicker, the maria occur only as small patches in the deepest basins and craters, such as Tsiolkovsky.

Mare flooding of some basins, such as Imbrium and Orientale, appears to have taken place in two stages. Partial filling by impact melt was followed by sporadic emplacement of mare basalts from both flood-type and plains-type eruptions. On the lunar nearside, the inner rings of most basins like Imbrium were completely or nearly completely inundated by lavas, and in many cases the outermost ring was breached, spilling lava from one basin to another. Basins in the limb regions like Orientale were only partly filled, if at all, and show an asymmetric distribution of mare toward the near side.

Irregular maria, such as Procellarum, and certain circular maria, such as Australe and Smythii, developed in a much more complex fashion. Smaller impact craters (less than 150 kilometers in diameter) were modified by tectonic and local volcanic processes that may have contributed to their inundation with mare materials.

Mapping the Surface of the Moon

Lunar geological mapping provides an important framework for nearly all geological and geophysical investigations and has allowed a better understanding of lunar history to be formulated. The entire nearside of the Moon has been mapped at scales of both 1:1,000,000 and 1:5,000,000. Mapping of the

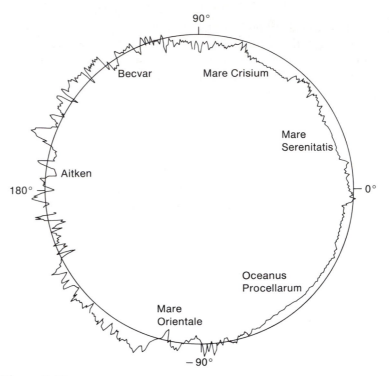

Figure 5.12
Radial plot of radar sounding profiles about the lunar center of mass, acquired
during the Apollo missions. The radius of the sphere is 1738 kilometers.
[Courtesy of Michael Kobrick, 1976.]

remainder of the Moon at 1:5,000,000 is still in progress, along
with selected detailed 1:250,000 mapping using high-quality
photography obtained from Apollo orbiters.

A synthesis of the inferred geological history is shown in the
generalized geologic time scale for the Moon (Fig. 5.13) and in
a series of pictures (Fig. 5.14) showing three stages in the evolution of the Moon, beginning with the formation of the Imbrium Basin and the deposition of its ejecta as the Fra Mauro
Formation. This event came close to the end of the period of

Figure 5.13 (*facing page*)
This lunar stratigraphic column indicates the generalized geological time scale for the
Moon. [From *Geology on the Moon* by J. E. Guest and R. Greeley, fig. 1.5. Taylor and
Francis, Ltd., London, 1977.]

HISTORY OF THE MOON

Time-stratigraphic Units	Date (years)	Rock Units	Events	Notes
Copernican System		Few large craters	Tycho Aristarchus	Craters with bright rays and sharp features at all resolutions (e.g., Tycho, Aristarchus)
		Few large craters	Copernicus	Craters with bright rays and sharp features but now subdued at meter resolutions (e.g., Copernicus)
Eratosthenian System		? Few large craters	Eratosthenes	Craters with Copernican form, but rays barely visible or absent
	3.2×10^9	Apollo 12 lavas		
	3.3×10^9	Apollo 15 lavas	Imbrium lavas	Few lavas with relatively fresh surfaces
Imbrian System	3.42×10^9	Luna 16 lavas		
			Few large craters	
		Mare lavas	Eruption of widespread lava sheets on nearside; few eruptions on farside	Extensive piles of basaltic lava sheets with some intercalated impact crater ejecta sheets
	3.6×10^9	Apollo 11 lavas		
	3.8×10^9	Apollo 17 lavas		
		Cayley Formation? / Hevelius Fm.	Orientale Basin	
	3.9×10^9	Fra Mauro Fm.	Imbrium Basin	
Nectarian System			Crisium Muscoviense Humorum Nectaris	Numerous overlapping, large, impact craters and associated ejecta sheets together with large basin ejecta
		Janssen Fm.	Serenitatis Smythii Tranquillitatis Nubium (Basins)	Any igneous activity at surface obscured by impact craters
	4.1×10^9			
Pre–Nectarian			Formation of Moon	"Crystalline" rocks formed by early igneous activity
	4.6×10^9			

Figure 5.14
The illustrations here and on the next two pages are artist's renderings of the face of the Moon at three stages in its history. Above is the present Moon. [From "The Two Former Faces of the Moon" by Don E. Wilhelms and Donald E. Davis, *Icarus*, vol. 15, fig. 1, 1971. Academic Press, Inc.]

Reconstruction showing the Moon at the end of the Imbrium Period, soon after the formation of most of the mare material approximately 3.3 billion years ago. Note the absence of such young craters as Tycho, Copernicus, and Eratosthenes from the otherwise not too unfamiliar scene. [From "The Two Former Faces of the Moon" by Don E. Wilhelms and Donald E. Davis, *Icarus*, vol. 15, fig. 2, 1971. Academic Press, Inc.]

Figure 5.14 *(continued)*
Reconstruction showing the Moon in the middle of the Imbrium Period, before the formation of the present mare surface material and after the formation of the last of the mare
basins, Mare Orientale (partly in view on the west limb). Features "exhumed" by the
artist include the fully developed, yet unburied Iridum crater, perched on one of the rings
of the Imbrium Basin. [From "The Two Former Faces of the Moon" by Don E. Wilhelms
and Donald E. Davis, *Icarus,* vol. 15, fig. 3, 1971. Academic Press, Inc.]

heavy bombardment and was followed by a period of sporadic basaltic eruption lasting probably more than a billion years. Continued impact cratering at a low rate peppered the surface of the Moon, degrading its primary surface features. The incompleteness of this process during the past three billion years, however, is indicated by the preservation of 10-meter features on lava flows of that age. Internal activity is evidenced by linear grabens and other fault structures and continues today on a minor scale, as revealed by deep moonquakes recorded by seismometers left by the Apollo astronauts.

THE INCOMPLETE EXPLORATION

Major Gaps in Knowledge

Why should we consider further lunar exploration after the abundant yield of the Apollo program? The most important reason is that the primary focus of Apollo was the landing of a man on the Moon, which resulted in very uneven geographic coverage of the Moon. The scientific activity and data collection were confined, by necessity, to those areas easiest and safest to reach for a manned landing vehicle from Earth—to nearside equatorial areas. The most distant point from the lunar equator that was sampled directly by Apollo is only 26° north latitude. The targets of Ranger, Surveyor, and Lunar Orbiter missions also were primarily in areas under consideration as eventual Apollo landing sites. Even the Apollo command module was placed in an equatorial lunar orbit to facilitate rendezvous with the lunar excursion model. Related technical limitations have narrowed the coverage of the Soviet Luna automated sample-collection missions.

As a result, there is great disparity between observational coverage of the nearside equatorial regions, which have been well studied, and the polar and farside regions, which have not. In fact, no systematic global reconnaissance of the Moon has ever been made. Thus, even though present scientific understanding of the Moon is the consequence of a tremendous engineering effort in response to the public enthusiasm for placing a man on the Moon, there remains, incongruously, an uneven distribution of sampling and exploration. Therefore, the first priority for future lunar exploration should be to fill in the gaps of observational coverage of the lunar surface, with emphasis on polar areas and on the lunar farside.

Furthermore, the enormous amount of scientific data acquired at great cost through the collection of lunar samples and other data from a few isolated localities may be significantly enhanced by relatively inexpensive new global measurements systematically acquired from a suitably instrumented polar orbiter.

For example, exhaustive laboratory analyses of lunar samples has defined three principal kinds of bedrock to constitute the lunar surface: (1) calcium-rich anorthosites, perhaps representing part of the primordial crust of the Moon; (2) younger mare basalts; and (3) highly radioactive KREEP basalts whose origin remains uncertain. It is very difficult to determine the distribution of these rocks accurately over the entire lunar surface or to recognize other constituents from the existing data. A properly instrumented satellite in lunar polar orbit (Fig. 5.15) could extend the detailed knowledge of the Apollo landing sites to the entire surface of the Moon. In addition, gravimetry and altimetry measurements provided by such a polar orbiter could pursue the question of variations in crustal thickness, especially the asymmetric distribution of crust between nearside and farside hemispheres.

Regional correlations of rock type and physiography could be obtained from uniform global observations. In particular, it may be possible to relate the chemical rock types to tectonic environment, as on the Earth. Perhaps an answer may be found to the unresolved question about a possible connection between the origin of mare basalt flows and any long-term effects persisting from earlier basin-forming impact events. In any case, additional high-resolution imaging acquired uniformly from a polar orbiter would permit much more accurate determination of relative ages of the various basalt flows, which could lead to a much more meaningful and detailed thermal history of the Moon.

Figure 5.15
Artist's view: Lunar Polar Orbiter. This possible future mission could provide a wide range of remote sensing data of the Moon that would give the first uniform global view permitting the surface truth data of the Apollo missions to be extrapolated over the entire surface.

Another unresolved problem left from the Apollo era is whether a lunar core exists, and, if so, what its state and composition might be. A detailed electromagnetic probe of the Moon by one or two orbiters would probably resolve the question of the existence of a core. The perplexing remanent magnetization found to be an intrinsic part of the lunar samples is also a clear target for further investigation.

An additional reason that further lunar research is of major scientific importance is that the Moon is the "calibration" planet for systematizing and comparing the observations of the other Earthlike planets, from which no samples have yet been returned, and are not likely to be for decades.

United States–Soviet Lunar Collaboration?

Even a rather modest lunar polar orbiter equipped with the ever-improving scientific instrumentation and data-handling capabilities would have the potential for a great yield of scientific information. At present, however, lunar science and exploration by the United States have been caught in a painful "valley" caused by the inevitable decline following the enormous accomplishment and cost of Apollo. This has brought to a virtual standstill acquisition of new lunar data by the United States, with only the prospect of renewed capability perhaps 10 or 15 years hence.

Until recently the Soviet Union continued its Luna series of automated rovers ("Lunakhod") and sample-collecting missions, as well as orbiters, at the rate of about one flight attempt per year. Soviet plans for future space activities, as described in the popular press, are very ambitious and include not only Earth-orbiting space stations, but also lunar-orbiting space stations and surface operations. However, actual progress has been quite slow. Soviet space scientists once had plans to develop an advanced Luna series capable of performng automated operations on the lunar farside, including automated rovers for collecting lunar samples from carefully chosen sites and delivering them to automated sample carriers for transportation to Earth. So far, however, there have been no signs of major new lunar developments despite the enthusiasm of Soviet planners.

One possible new dimension of future lunar scientific exploration is the potential for joint operations by the Soviet Union and the United States. To some extent, the Soviet proven automated transportation systems—soft-landers, rovers, and sample-collection vehicles—and the United States' capability in space instrumentation and inflight and ground computing are complementary. Thus joint farside and polar automated missions, perhaps utilizing relay satellites, are within the realm of technical, if not political, feasibility.

SUGGESTED READING

Anon. *Atlas and Gazetteer of the Near Side of the Moon.* NASA SP-241. 1970.

Anon. DOD Catalog of Aeronautical Charts and Flight Information Publications. U.S. Air Force, 1969.

Arthur, D. W. G., and E. A. Whitaker (compilers). *Orthographic Atlas of the Moon—Supplement Number One to the Photographic Lunar Atlas.* NASA NsG-37-60 and Contract AF-19(604)-7260, Aeronaut. Chart Inform. Center, 1960.

Bowker, D. E., and J. K. Hughes. *Lunar Orbiter Photographic Atlas of the Moon.* NASA SP-206. Washington, D.C.: U.S. Government Printing Office, 675 plates, 1971.

Burnett, D. S. "Lunar Science: The Lunar Legacy." *Rev. Geophys. Space Phys.* **13**:13–34, 1975.

El-Baz, F. "The Moon After Apollo." *Icarus* **25**:495–537, 1975.

Guest, J. E., and R. Greeley. *Geology on the Moon.* London: Wykeham Publications, 1977.

Hansen, T. P. *Guide to Lunar Orbiter Photographs.* NASA SP-242, 1970.

Head, J. W. "Lunar Volcanism in Space and Time." *Rev. Geophys. Space Phys.* **14**:265–300, 1976.

Kopal, Z. (ed.) *Physics and Astronomy of the Moon.* New York: Academic Press, 1961.

Kopal, Z., J. Klepesta, and T. W. Rackman. *Photographic Atlas of the Moon.* New York: Academic Press, 1965.

Kosofsky, L. J., and F. El-Baz. *The Moon as Viewed by Lunar Orbiter.* NASA SP-200, 1970.

Kuiper, G. P., D. W. G. Arthur, E. Moore, J. W. Tapscott, and E. A. Whitaker. *Photographic Lunar Atlas: Based on Photo-*

graphs Taken at the Mount Wilson, Lick, Pic du Midi. McDonald and Yerkes Observatories. Nat. Sci. Foundation (Contrct AF-19(604)-3873), 1960.

Manned Spacecraft Center. *Apollo 11 Preliminary Science Report.* NASA SP-214. Washington, D.C.: U.S. Government Printing Office, 1969.

Manned Spacecraft Center. *Apollo 12 Preliminary Science Report.* NASA SP-235. Washington, D.C.: U.S. Government Printing Office, 1970.

Manned Spacecraft Center. *Apollo 14 Preliminary Science Report.* NASA SP-272. Washington, D.C.: U.S. Government Printing Office, 1971.

Manned Spacecraft Center. *Apollo 15 Preliminary Science Report.* NASA SP-289. Washington, D.C.: U.S. Government Printing Office, 1972.

Manned Spacecraft Center. *Apollo 16 Preliminary Science Report.* NASA SP-315. Washington, D.C.: U.S. Government Printing Office, 1972.

Manned Spacecraft Center. *Apollo 17 Preliminary Science Report.* NASA SP-330. Washington, D.C.: U.S. Government Printing Office, 1973.

Moore, P. *A Survey of the Moon.* W. W. Norton, 1963.

Mutch, T. A. *Geology of the Moon.* Princeton: Princeton Univ. Press, 1972.

Proceedings of the Apollo 11 Lunar Science Conference. *Geochimica et Cosmochimica Acta,* Supplement 1 (3 volumes). New York: Pergamon Press, 1970.

Proceedings of the Second Lunar Science Conference. *Geochimica et Cosmochimica Acta,* Supplement 2 (3 volumes). Cambridge: MIT Press, 1971.

Proceedings of the Third Lunar Science Conference. *Geochimica et Cosmochimica Acta,* Supplement 3 (3 volumes). Cambridge: MIT Press, 1972.

Proceedings of the Fourth Lunar Science Conference. *Geochimica et Cosmochimica Acta,* Supplement 4 (3 volumes). New York: Pergamon Press, 1973.

Proceedings of the Fifth Lunar Science Conference. *Geochimica et Cosmochimica Acta,* Supplement 5 (3 volumes). New York: Pergamon Press, 1974.

Proceedings of the Sixth Lunar Science Conference. *Geochimica et Cosmochimica Acta,* Supplement 6 (3 volumes). New York: Pergamon Press, 1975.

Proceedings of the Seventh Lunar Science Conference. *Geochimica et Cosmochimica Acta,* Supplement 7 (3 volumes). New York: Pergamon Press, 1976.

Proceedings of the Eighth Lunar Science Conference. *Geochimica et Cosmochimica Acta,* Supplement 8 (3 volumes). New York: Pergamon Press, 1977.

Proceedings of the Ninth Lunar Science Conference. *Geochimica et Cosmochimica Acta,* Supplement 10 (3 volumes). New York: Pergamon Press, 1978.

Proceedings of the Soviet-American Conference on the Geochemistry of the Moon and Planets. NASA SP-370 Washington, D.C.: U.S. Government Printing Office, 1977.

Proceedings of the Tenth Lunar Science Conference. *Geochimica et Cosmochimica Acta,* Supplement 11 (3 volumes). New York: Pergamon Press, 1979.

Special Moon Issue. *Science* **167**(3918):449–782, 1970.

Taylor, S. R. *Lunar Science: A Post Apollo View.* New York: Pergamon Press, 1975.

Whitaker, E. A., et al. *Rectified Lunar Atlas—Supplement Number Two to the Photographic Lunar Atlas.* Aeronaut. Chart Inform. Center, 1963.

Note: The U.S. Geological Survey has published a series of geological maps of the entire Moon at a scale of 1:5 million, and of the nearside at a scale of 1:1 million as part of its Miscellaneous Investigations Map Series.

6

THE RECORD FROM MERCURY

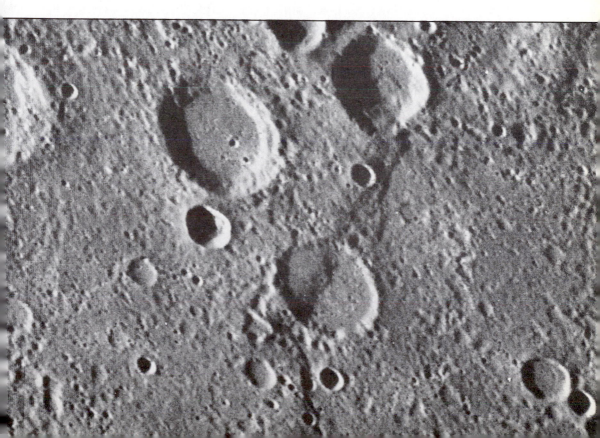

THE RECORD
FROM MERCURY

Mercury was hardly known, even as an astronomical object, until Mariner 10, the first and only mission to Mercury, flew by it in 1974 and returned detailed pictures of its surface. Observations from Earth are intrinsically difficult because Mercury is the innermost planet in the Solar System and never strays more than 28° away from the Sun as seen from Earth. As a consequence, before 1974, Mercury was practically ignored by scientists—even by science fiction writers!

What little was known about the planet was puzzling. Paradoxically, Mercury appeared to be like the Moon on the outside and the Earth on the inside. Mercury closely resembles the Moon in the way it reflects sunlight and radar pulses, and also in the way that it emits heat radiation at infrared and radio wavelengths. These resemblances mean that, like the Moon, Mercury must be covered with at least a thin layer of dry, fine-grained, dark silicate material. Yet internally the planet is much denser than the Moon and even rivals the Earth in that regard.

The gain in information about Mercury by Mariner 10, compared with previous Earth-based measurements, was extraordinary. The features evident in the Mariner 10 photography and the concurrent discovery of Mercury's small, but apparently Earthlike magnetic field have elevated that paradox to a question of great importance in the study of the inner planets. The similarity to the Moon in surface reflectance and emission of electromagnetic waves was found to accompany a much more detailed similarity to the lunar surface—both in the kinds of topographic features present on Mercury and in their relative sequence of ages. Despite the fact that the two bodies are in very different parts of the Solar System, where the nature and frequency of impacting objects might be quite different, they apparently experienced quite similar histories of impact bombardment.

THE EARLIEST TESTIMONY—HEAVILY CRATERED TERRAINS AND INTERCRATER PLAINS

The early history of Mercury, like that of the Moon, is hidden from easy view by numerous landforms generated by later impact and perhaps other less well understood processes. Large, old craters and the areas between them (called "intercrater plains") dominate most of the mercurian landscape, which appears superficially similar to the lunar highlands.

Chemical Separation of the Planet-Forming Materials

Mercury differentiated chemically early in its history. This is indicated by several lines of evidence. First, the planet's very high bulk density and rocky exterior is consistent with an outer covering of silicate material, at most 500–600 kilometers thick, lying atop a dense, iron-rich core (Fig. 6.1). The interpretation that volcanism is the source of many of the plains seen in Mariner 10 photography adds to the plausibility that Mercury is differentiated, since volcanism would require the outer silicate layer to be at least several hundred kilometers thick in order to provide a source for the lava. Finally, even more compelling as evidence for a differentiated Mercury is its Earthlike magnetic

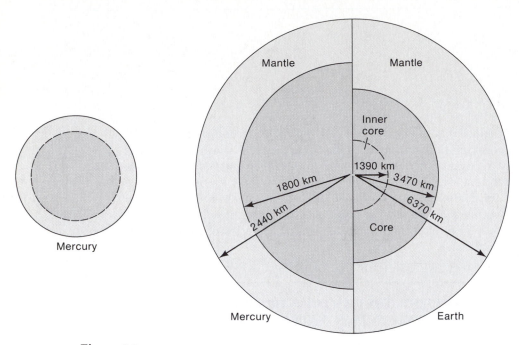

Figure 6.1

A comparison of the internal constitutions of Mercury and Earth. These cutaway views are scaled to the same outside diameter in order to show how much larger Mercury's iron core is thought to be compared with Earth's core. The cores are shown in dark gray. Mercury's actual size in relation to the Earth is shown by the small disk at the left. Mercury's iron core evidently contains 80 percent of the planet's mass. Therefore, the iron core must have a radius of at least 1800 kilometers, which would make the core alone slightly larger than the Earth's Moon. [From "Mercury" by B. C. Murray. Copyright © 1975 by Scientific American, Inc. All rights reserved.]

field. At the least, a large iron core seems required; a fluid, convecting iron core may be indicated as well.

Like the Moon and the Earth, Mercury must have differentiated and cooled very early for a rigid lithosphere to have formed before the formation of the oldest terrains that are presently extant. Similarly, there is no evidence of the action on the surface of any early atmosphere. Hence devolatilization, if any, must have been completed before the present surface was formed. Mercury probably passed through a very active thermal regime early in its history and may well have been entirely molten at the end of accretion.

Degradation and the Early History of Mercury

Two aspects of the mercurian landscape testify to one or more periods of intense cratering and resurfacing: areas of closely spaced large craters and large tracts of intercrater plains (Fig. 6.2). In many regions, the plains materials cannot be seen to lap onto the craters, so that the superposition of secondary craters onto the plains can be taken as evidence that the large craters there postdate the formation of the plains. In other regions, however, the subdued topography revealed in stereoscopic images suggests that some portions of intercrater plains do overlap cratered terrain (Fig. 6.3), suggesting that plains formation took place after crater formation. In both cases, the existence of the plains indicates that extensive resurfacing of Mercury occurred soon after accretion, when the surface of Mercury was bombarded repeatedly by large crater-forming objects. Intercrater plains on the Moon are not as extensive as those on Mercury.

What was this ancient degradational process that resurfaced the intercrater plains after the end of accretion? Three possibilities can be considered: (1) eolian, (2) ballistic, and (3) volcanic. The arguments against eolian phenomena on Mercury arise by analogy with Mars, where the presently observed 6-millibar atmosphere has been more than sufficient for surface winds to blur morphologic features of low relief (such as rays), which remain undisturbed on Mercury. It must be concluded that no tangible atmosphere has affected the ephemeral surface features of Mercury.

Two observations suggest that ballistic processes (i.e., the impacts of meteorites and crater ejecta) are not responsible for the landform degradation. First, for a given crater diameter, the areal extent of crater ejecta is more limited on Mercury than on the Moon because of the greater gravitational acceleration (Fig. 6.4). Second, the total number of large craters per unit area appears to be less on Mercury than on the Moon (Fig. 8.1). Thus the combination of the two effects should have significantly *reduced* rather than increased the effectiveness of erosion and deposition by impact-generated ballistic processes as compared to the Moon.

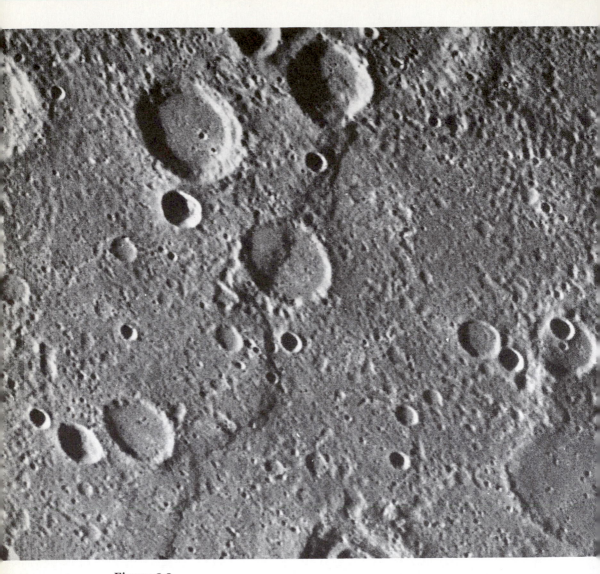

Figure 6.2
Intercrater plains and heavily cratered terrain are typical of much of Mercury outside the area affected by the formation of the Caloris Basin. **Above.** Abundant, shallow elongate craters and crater chains on the Santa Maria Plains, a large tract of intercrater plains (centered at 3°N, 20°W). A prominent scarp, Santa Maria Rupes, cuts both intercrater plains and old craters. Area shown is 200 kilometers across. In all views north is at the top. **Right.** Low-resolution view of the northern plains. Large craters and small basins can be recognized that are similar to the heavily cratered terrain on the Moon. The Moon's intercrater plains are not generally recognized at comparable resolution. Roman numerals in the outline drawing indicate craters of various morphologic classes, from fresh (I) to highly degraded (V). Letters (A to C) indicate regions of intercrater plains that display roughly craterlike, planimetric form.

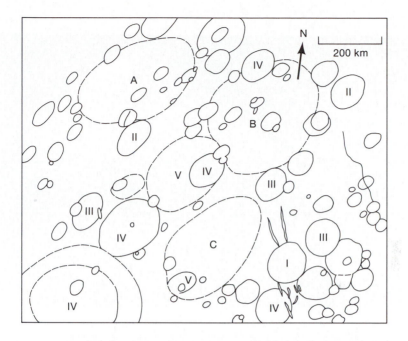

Figure 6.3

The superposition of intercrater plains is illustrated in these sets of Mariner 10 frames rectified to orthographic projection at the Image Processing Laboratory, Jet Propulsion Laboratory. **Above.** Frames centered near 43°S, 45°W. Sun elevation is approximately 25°. Arrows in the upper part of the area shown denote the periphery of a large, craterlike depression comprising an area of intercrater plains. Smaller crater forms (A) embay the rim deposits of a large, well-defined basin (B), but are themselves modified portions of the intercrater plain. **Upper right.** Stereoscopic pair of Mariner 10 frames centered near 22°S, 50°W. Sun elevation angles near 40°. Arrows denote textural boundary: To the north, the concentric rings of basin A are well defined morphologically; to the south, almost no trace of the multiring structure is visible. However, in stereoscopic viewing, the topographic manifestation of the rings is continuous across the boundary. Note the large escarpments near B and the linear trend A–B. **Lower right.** Stereoscopic pair of Mariner 10 photographs centered near 65°S, 50°W, with the Sun about 17°. The letter A denotes a large ridge that crosses the large escarpment marked B. Arrows denote the approximate location of a lobate front of intercrater plains material that transects several crater walls without significantly deforming the craters. This can be interpreted as evidence of volcanic flows within the intercrater plains.

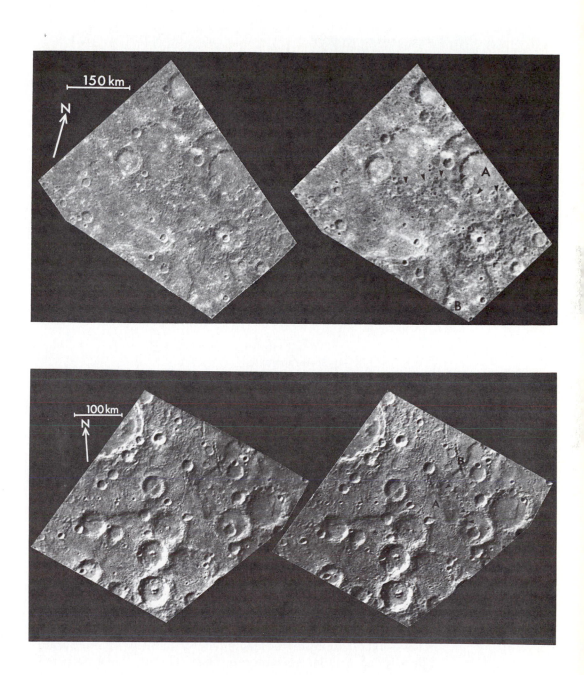

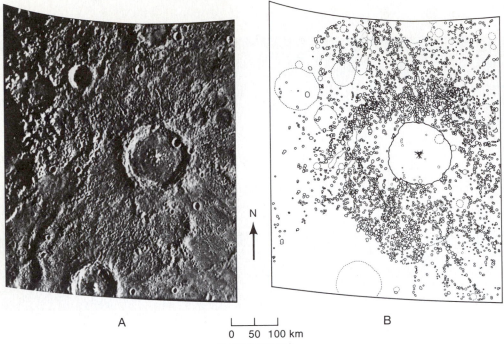

A   N ↑   B

0 50 100 km
Scale (approx.)

Figure 6.4
Left. This orthographic projection of a Mariner 10 image shows a 140-kilometer-diameter crater and its surrounding zone of secondary craters. The narrow width of the rim facies, the prominent subradial secondary crater chains, and grooves are representative of the larger mercurian craters. **Right.** The field of secondary craters surrounding the crater is depicted as mapped from an enlarged rectified image.

The preceding arguments leave volcanic processes as the most plausible mechanism for ancient landform degradation, although there is no irrefutable evidence that volcanism has ever occurred on Mercury. Overlap relationships in some areas show large, lobate scarps bounding regions of intercrater plains; these suggest enormous flows of intercrater plains material. Volcanism is a plausible—but nevertheless mainly *ad hoc*—explanation for the degraded nature of early mercurian landforms. Similar volcanic episodes also may have occurred in the Moon's early history, but the landforms they produced would have been obliterated by the later bombardment of meteorites whose debris blanketed older terrains over greater distances than on Mercury.

Chronology of the Early History of Mercury

The most controversial subject concerning the early history of Mercury, like that of the Moon, is the chronology of events: When did the massive cratering occur? Did it end quickly?

The controversy has been carried to Mercury through interpretation of the Mariner 10 photographs. Without samples of the surface materials of Mercury, the absolute chronology cannot be determined. However, the predominance of intercrater plains on Mercury, as compared to the Moon, may permit a look slightly further back into the planet's early sequence of events.

Two principal hypotheses concerning this early history are that either an extended period of obliteration was followed by a discrete episode of bombardment or crater formation and the obliteration of craters by plains formation took place continuously and simultaneously.

The first hypothesis is developed in the following way. The similarity between lunar and mercurian degraded craters (and their lack of similarity to martian craters) means that melting, crustal plasticity, or atmospheric interaction have not modified the appearance of these Moonlike features. Yet the intercrater plains in some areas clearly are older than the large craters. All craters formed during the accretion of the planet were apparently completely erased before the present craters formed.

Hence it is inferred that the large craters must have formed in an older host material (although in a few very heavily cratered areas, craters may survive that predate the plains in those areas). Alternative ways to explain resurfacing, including "cataclysmic resurfacing" by basin ejecta, are deemed unlikely on the grounds that the ejecta would not cover enough area to produce such widespread intercrater plains.

Thus an extended period of obliteration followed by a discrete episode of bombardment, correlated with a cataclysmic lunar bombardment, is one simple explanation of the picture data.

The second hypothesis differs principally by postulating continuous crater obliteration by volcanism rather than a terminal episode of bombardment. In this case, volcanism is inferred to have competed with continuous impact bombardment and then ended at about the time the heavy bombardment ceased. Whether a terminal episode of bombardment is required to explain the Mariner 10 photographic observations is a question that is likely to remain open for some time.

Interplanetary Correlations

Comparisons of mercurian surface features with those on Mars and the Moon have shown four interesting correlations. (1) The size-frequency distributions of craters and basins on all three planets are similar in slope (Fig. 6.11). This similarity supports the idea that bombardment ceased at the same time everywhere in the inner Solar System. (2) The abundance of craters greater than 10 kilometers in diameter, when corrected for variations in impact/encounter velocities, is roughly similar for each planet. This observation, too, suggests to some scientists a synchronous bombardment history. (3) Each planet has roughly the same number of large craters, even though the planets differ in area by a factor of four. This observation remains unexplained. (4) The presence of intercrater plains on all three planets has revived the controversy over the origin of the many types of highland plains on the Moon. As discussed in Chapter 5, some of these plains were originally thought to be volcanic, but analyses of Apollo samples tend to support an impact-generated origin. (Other old intercrater plains exist in small patches on the Moon as well.) With the larger expanses of similar plains on Mercury and on Mars—and with evidence there suggesting volcanism—this point remains one of the more controversial aspects of interplanetary comparison.

Tectonics and Scarps

The final event in the early history of Mercury was the formation of the large escarpments. However, both their modes and times of origin are far from clearly discerned. Scarps are observed to cut certain large craters and not others, indicating that most scarps formed near the end of the period of heavy bombardment. In addition, another variety of scarps formed later to cut much younger plains units, but these are not clearly of a global compressional nature. A few old scarps displace portions of crater rims, suggesting compressive (thrust) fault movement—a feature not recognized significantly on the Moon or Mars. Tectonic trends have been mapped, and it has been inferred that the crust of Mercury may have contracted globally to produce compressional scarps, perhaps through interaction with the global stresses produced in the planet as its spin was gradually slowed by the tidal (gravity) effects of the Sun. This contraction may, in turn, be related to the thermal history of the planet's interior, when (and if) the large core shrank slightly as it cooled.

Thus the early history of Mercury seems complex in its diversity of phenomena. High rates of bombardment and possible volcanic resurfacing, either episodically or continuously, produced huge tracts of cratered terrain and intercrater plains. The formation of scarps and plains may have been integrally related to the thermal and dynamic history of the evolving planet. Heavy bombardment, whether continuous or episodic, ended relatively abruptly, as on the Moon.

THE CALORIS EVENT

The most spectacular surface feature viewed by Mariner 10 during its three encounters with Mercury is the circular impact basin Caloris (Fig. 6.5). Although less than half of the basin was illuminated during the Mariner 10 flybys, enough was seen to reveal its structure for comparison with circular basins on other planets and to pique the interest of planetologists.

The Caloris Basin is defined by a ring of mountains about 1300 kilometers in diameter. The ring is composed of smoothly rounded blocks rising 1–2 kilometers above the surrounding terrain. The basin interior is filled with ridged plains units that are highly fractured, hinting at tectonic deformation and giving the basin a unique character.

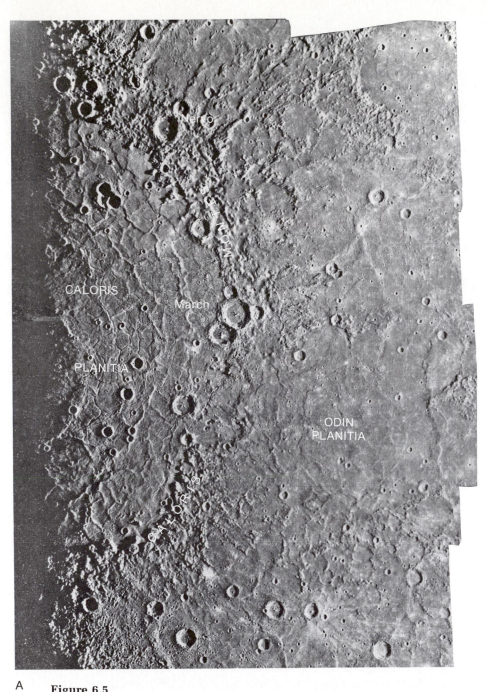

A

Figure 6.5
(A) This mosaic of the Caloris Basin prepared from Mariner 10 images, shows named features. The Caloris Montes define the basin diameter at about 1300 kilometers.

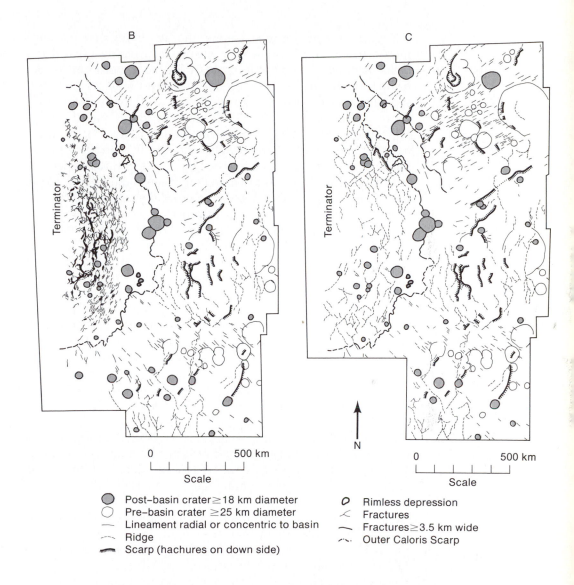

Post–basin crater ≥ 18 km diameter
Pre–basin crater ≥ 25 km diameter
Lineament radial or concentric to basin
Ridge
Scarp (hachures on down side)

Rimless depression
Fractures
Fractures ≥ 3.5 km wide
Outer Caloris Scarp

(B) Fractures and other features associated with the Caloris Basin are outlined in this diagram.

(C) Diagram showing ridges and associated features. [From "Tectonism and Volcanism on Mercury" by Robert Strom, Newell Trask, and John Guest, *Journal of Geophysical Research*, vol. 80, fig. 2, p. 2480, June 10, 1975. Copyrighted by the American Geophysical Union.]

Typical of most circular basins, the mountain ring is asymmetric in cross section with a steep scarp facing toward the basin's center. The mountain chain (in the eastern half of the basin, which was illuminated at the time of Mariner 10's passages) is nearly continuous except for a prominent discontinuity that forms a gap about 300 kilometers long. Smooth plains appear to embay this gap from outside the basin. Within parts of the mountain chain are valleys that are floored with plains material. These valleys may be graben structures. Like the mountain rings of circular basins on other planets, the Caloris Mountains are probably blocks of crust that were uplifted as part of the basin-forming impact process.

The Caloris Basin does not display all of the features of multi-ringed basins seen on the Moon. Only the Caloris Mountains are well defined. Northeast of these mountains, however, at a radial distance of about 130 kilometers farther out, there is an ill-defined scarp that is at least 800 kilometers long and concentric to the basin. The northern end of the scarp disappears into the darkness beyond the terminator, so that its total length is unknown. This scarp could be the manifestation of incipient block-faulting associated with basin deformation.

Caloris Basin Deposits

Surrounding the Caloris Mountains in irregular patches are areas called the Caloris lineated terrain, characterized by long, hilly ridges interrupted by radial grooves that cut pre-Caloris craters. This terrain can be traced for about 1000 kilometers, or nearly one basin diameter from the rim (Fig. 6.6). Morphologically, it closely resembles the lunar Imbrium sculptured terrain and is also similar to parts of the lunar Fra Mauro Formation. On the basis of analogous morphology and association, the Caloris lineated terrain is interpreted to consist of basin-continuous deposits. Such a unit may serve as a good time-stratigraphic horizon for geological mapping on Mercury, much as the Fra Mauro Formation serves as the base of the Imbrium System on the Moon.

Superimposed on the Caloris lineated terrain, and embaying the "gap" of the Caloris Mountains, are smooth and hummocky

Figure 6.6
This view of Caloris sculpture in the northeastern quadrant of the Caloris Basin shows the domical terrain (A) between the inner and outer scarps and the well-developed radial system (B) east of the outer scarp. Part of the interior of the Caloris Basin is seen at (C).

terrain units concentrated in Suisei Planitia northeast of the basin, Odin Planitia to the east, and Tir Planitia to the southeast (Fig. 6.7). (Odin and Tir are the names of Mercury in the ancient Norse and ancient Persian languages, and Suisei is its name in the modern Japanese language.) Two types of plains units can be distinguished in this terrain—smooth plains and hummocky plains. Both are cut by mare-type ridges. Hummocky plains occur immediately outside the Caloris Mountain ring. This unit is composed of low, closely spaced to scattered hills

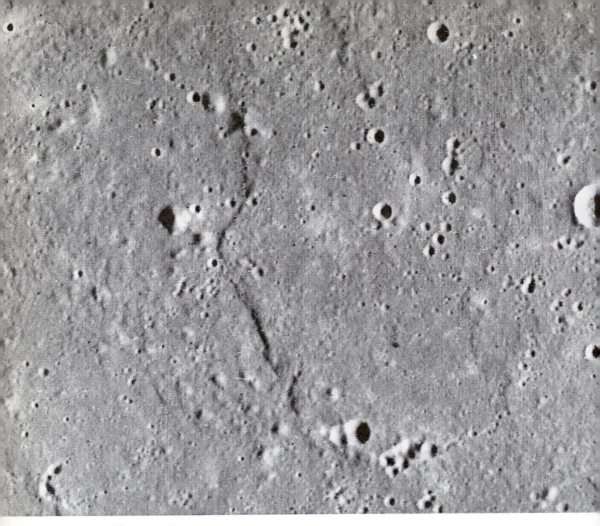

Figure 6.7
Smooth-to-hummocky plains east of the Caloris Basin are shown in this Mariner 10 image. The scarp near the center of the picture is about 350 meters high and may be a fault or possibly a flow front of partially melted Caloris ejecta, as discussed in the text. The large hill left of the scarp is about 1.4 kilometers high. The area covered by this picture is 153 by 115 kilometers. Illumination is from the east (right).

that are about 0.3–1 kilometer across. Some of the hills are arranged in concentric bands around the basin, giving the unit a corrugated appearance somewhat like the domical and corrugated units (Montes Rook Formation) surrounding the Orientale Basin on the Moon (Chapter 5). The hummocky plains material of Caloris may be basin ejecta composed of a mixture of unmelted fragments and impact melt. Some areas of the hum-

Figure 6.8
Part of the floor of the Caloris Basin exhibiting characteristic ridges and fractures. Notice that the length and width of the fractures increase toward the center of the basin (lower right to upper left) and that the fractures transect the ridges at various angles. Illumination is from the east (right).

mocky plains display short, inward-facing scarps that have been speculated to be flow fronts produced by the basinward flow of impact melt shortly after impact.

Basin Deformation

Smooth plains filling the Caloris Basin have prominent ridges and fractures (Fig. 6.8) whose patterns and orientations record post-basin-filling deformation. Ridges on the floor of the basin are 1.5–12 kilometers wide and appear to be about 500–700 meters high on the basis of shadow lengths and brightness variations. They form an approximately polygonal pattern, elements of which are concentric and radial to the general orientation of the basin. The ridges are likely the result of horizontal compression radially oriented with respect to the center of the basin.

A well-defined pattern of fractures is also displayed on the smooth plains within the Caloris Basin. Fractures frequently cut ridges, and some occur along ridge crests, showing that the fractures postdate the ridges. Fractures range in width from about 8 kilometers down to the limits of resolution (about 700 meters). The larger fractures can be discerned to be grabenlike, with flat floors as much as 700 meters deep. If these large fractures are grabens, tensional processes in the Caloris floor are indicated subsequent to the presumably compressional conditions associated with the ridges. Most of the fractures have a concentric trend, although some are radial.

These geometric and age relations within the Caloris Basin pose unresolved problems. Some sort of multistage deformational sequence is suggested after the emplacement of the Caloris smooth plains. It has been theorized that subsidence first accompanied extrusion of volcanic plains outside of Caloris and that later uplift occurred due to slow rebound from the impact event. Analogous deformational patterns have not been recognized on the Moon or Mars.

Possible Relation of Caloris to the Hilly and Lineated Terrain

Mariner 10 discovered an extensive region characterized by a peculiar hilly and jumbled surface (Fig. 6.9) and mapped as *hilly and lineated terrain*. The hills are about 5–10 kilometers wide and 0.1–1.8 kilometers high. Rims of craters in the area have been broken up into hills and depressions. The region is quite limited in area and clearly reflects a degradation process

Figure 6.9
Hilly and lineated terrain occur in a location approximately antipodal to the Caloris
Basin on Mercury. Large craters there show extensive modification. Several display numerous subdued wall furrows but exhibit preserved crested rim profiles. Plains units
occur within shallow, modified craters. Intercrater regions are hillocky, with transecting
sets of NE- and SW-trending furrows.

not evident elsewhere on the half of the mercurian surface so
far observed.

The hilly and lineated terrain is approximately antipodal to
the Caloris Basin. Somewhat similar terrains are observed on
the Moon (see Chapter 5) antipodal to the Imbrium and Orientale basins and have been attributed to a focusing of basin ejecta.
Because of the higher gravity and larger size of Mercury, however, basin ejecta would be expected to have fallen far short of

the antipodal regions. Calculations do suggest, however, that substantial surface deformation conceivably might take place as a result of the focusing of seismic waves generated by a basin-forming impact especially enhanced on Mercury by focusing of seismic energy through the large core. Figure 6.10 shows the preservation of both crested rim profiles and relatively small hills and pits. These features suggest a sudden catastrophic degradation and are inconsistent with long-term processes, such as extended impact erosion. The important observation in this respect is that only relatively young craters are unaffected. The rims of all old craters have been broken into hills and pits. Only on the margins of the terrain is there partial breakup. Impact erosion would not be expected to cause this type of local damage, since impact erosion should have affected the entire planet.

Comparisons with Basins on Other Planets

As each planet is explored, similarities and differences can be compared. Because the mercurian crust preserves a cratering record that, in many respects, is similar to that of the Moon, it would be surprising if no circular and multiringed basins had been found on Mercury. Nevertheless, Mercury's circular and multiringed basins do have features that are unique, especially the Caloris Basin.

First, Caloris does not show the well-developed multiringed character of the lunar Imbrium and Orientale basins, nor does it even show the multiringed character of the smaller mercurian basins. In this regard, it is more similar to the Hellas Basin of Mars. Of course, the lack of recognizable multiple rings within Caloris may merely be the result of subsequent burial beneath the smooth plains.

Second, Caloris' ejecta deposits differ from those of lunar basins. For example, there is no equivalent at Caloris of Orientale's braided and swirly ejecta. Perhaps this facies of the ejecta also has been covered by smooth plains, or perhaps it simply did not develop in the first place. Within the Caloris Basin, the terrain made up of small, closely spaced, domelike hills is in some respects similar to that formed in Orientale within the Rook Mountains, interpreted to be primarily impact melt. The Orientale terrain also displays a strong fracture pattern; the material of Caloris lacks these fractures.

Third, the biggest difference between the Caloris Basin and the basins seen on the Moon and Mars is in the distinctive fracture and ridge pattern of the floor. The ridges and the irregular, elongated depressions formed by the fractures suggest deformation associated with post-impact processes. In this regard, the basin-filling process seems to be more related to the basin-forming event than on the Moon. Perhaps Mercury's solid outer crust was thin enough at the time of the Caloris impact that immediate igneous response was possible, in contrast to the filling of the Imbrium Basin on the Moon, for example, where hundreds of millions of years transpired.

As on the Moon, basins play an important part in the fundamental crustal structure and stratigraphy of Mercury. An important part of future missions will be an effort to image the other half of the Caloris Basin as well as to determine whether other large circular basins exist on the half of the planet not seen by Mariner 10.

THE SMOOTH PLAINS OF MERCURY

About 15 percent of the observed hemisphere of Mercury is composed of smooth plains, generally similar in overall appearance to the lunar maria. Like the maria, these are among the youngest rock units on Mercury and, if they likewise are of volcanic origin, would suggest a remarkable similarity in the internal histories of the two planets.

Are the Smooth Plains Volcanic?

Although the smooth lunar maria exhibit distinctive features that are almost unambiguous indications of volcanic activity, no features of unambiguous volcanic origin are disclosed in Mariner 10 photographs of Mercury. This is not surprising in view of the limited coverage and resolution of the pictures. Furthermore, the planet was photographed at only one illumination condition. Nevertheless, the inference that most of the smooth mercurian plains are of volcanic origin must be based upon indirect arguments. As a consequence, the alternative has been proposed that the smooth plains of Mercury are not analogous in origin to the lunar maria but rather to the older light plains on the Moon (see Highland Plains in Chapter 5).

The smooth plains that fill the Caloris Basin exhibit unique fracturing and ridging not evident elsewhere on Mercury and clearly evidence controlling stresses related to the basin structure itself. In addition, the Caloris smooth plains exhibit about the same albedo as the surrounding material and show no evidence of originating at a different time from the surrounding mountains. Hence it cannot be ruled out that the Caloris plains may be the result of impact melt produced by the creation of the Caloris Basin itself rather than post-impact volcanic extrusions.

However, the impact-melt hypothesis for origin of the Caloris plains differs profoundly from any lunar example. The volume of the mercurian smooth plains is enormous—far too great to imagine that it could have originated from formation of the basin alone. On the Moon, impact melt seemingly constitutes a very small volume of the total impact excavation. In contrast, the Caloris Basin contains an enormous volume of fill. It overlaps and covers all other features that might have been associated with the impact itself. Therefore, if the Caloris plains are in fact of impact origin, they have resulted from an impact process in which a very much larger amount of impact melt was produced and retained than appears to have been the case for any lunar basin.

The annular band of plains within one basin radius may well contain much of the ejecta from the formation of the Caloris Basin. However, even within those plains, and for several more basin radii farther out, there are, in addition, very smooth plains. They are perhaps slightly younger and of very substantial volume compared to that which would have been produced by formation of the Caloris Basin itself. Because of Mercury's higher gravitational field, the ejection of material from the basin would be confined more laterally than on the Moon. Hence there would seem to be no analog on the Moon to support a nonvolcanic origin for the plains several radii away from Caloris Basin and peripheral to it.

In addition, the planet has other areas of plains (Fig. 6.10) that seem even more difficult to explain as of nonvolcanic origin. For example, in the northern area of the eastern hemisphere observed by Mariner 10, a 350-kilometer basin is asymmetrically filled by plains material of clearly younger age. The presence on the plains of an impact crater, probably created during formation of the plains, argues for an extended period of plains formation, presumably well separated from the time of basin formation. In other areas, there is clear-cut evidence of color differences between the plains and the surrounding older terrains. This argues against the derivation of the material by mass wasting from that surrounding terrain. Several distinct ages of plains have been noted as well, which is consistent with volcanic origin.

Finally, the Mariner 10 pictures show ample evidence of smooth plains material filling craters that clearly were created in a distinctly earlier event. Such occurrences would be difficult to explain on the basis of ejecta from other basins, as one would expect to see traces of this ejecta outside the craters as well. It is not possible to rule out nonvolcanic hypotheses completely without positive identification of unambiguous volcanic features in the photography, but the most probable interpretation at this point is a volcanic origin for much of the mercurian smooth plains.

A

Figure 6.10

(A) This computer-generated photomosaic of Mariner 10 images covers approximately 2.5×10^6 square kilometers of the north polar region of Mercury. At the top and bottom of the mosaic are dark, smooth plains interpreted to be volcanic. Note the ridges on these plains, which suggest post-formation deformation. The large craters, in particular the multiringed basins, show well-preserved and accented secondary crater fields that reflect the influence of gravitation on limiting the spread of ejecta as compared with the Moon. Enigmatic features in this area include the cluster of circular depressions at the lower left center, presumably impact craters modified by subsequent processes. **(B)** This computer photomosaic of Mariner 10 pictures shows the south polar region of Mercury. Longitude 90°W runs horizontally across the center of the figure, and longitudes 15°W and 185°W lie along the terminator. The lower half of the mosaic shows some of the youngest "intercrater plains" on Mercury. These plains, formed early in mercurian history, display large, arcuate ridges and scarps that likely reflect Mercury's unique history of planetary contraction and tidal despinning. The upper half of the figure shows more typical, older "intercrater plains" and "heavily cratered terrain." Large multiringed basins and one bright-rayed crater dominate the scene. Near the top of the frame is a bright spot—the only "mountain" yet seen on Mercury. It is interpreted to be an old, nearly unrecognizable volcano.

B

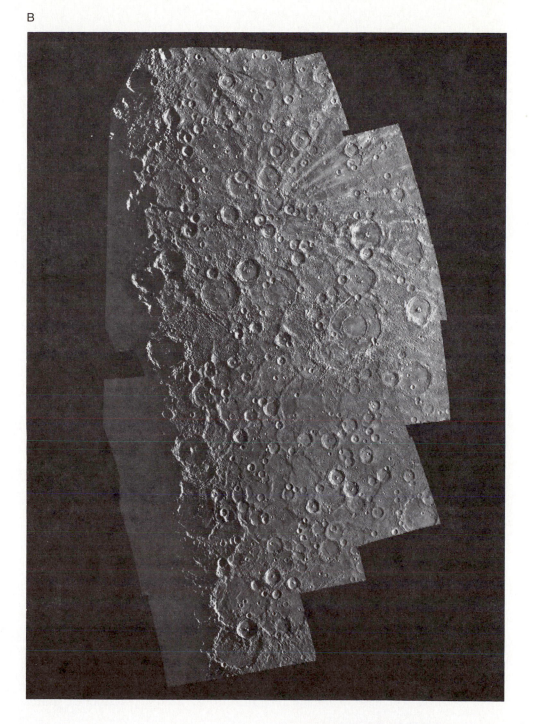

Similarity of Cratering on the Smooth Plains of the Moon, Mercury, and Mars

The smooth plains of the Moon, Mercury, and Mars exhibit a similar population of craters less than about 15 kilometers in diameter. These craters are characterized by a relatively fresh, bowl-shaped appearance, rather than a degraded one with a floor resulting from extensive landsliding. Most have well-developed rims. It has become increasingly apparent to specialists that these craters have experienced not only less degradation but probably a different kind from that which modified craters existing on the three planets before emplacement of the smooth plains. Just what those differences may be is the subject of active research. It is clear, however, that the smooth plains on the Moon, Mercury, and Mars record a similar population of impacting objects.

Figure 6.11 illustrates the conventional way in which crater populations are compared. From this size-frequency graph, it can be seen that crater frequencies obtained from Mariner 10 photographs of Mercury are close to those measured for the lunar maria, and also to the frequencies on the oldest of the smooth plains on Mars. These results were quite unexpected and are subject to at least two different interpretations.

One interpretation is that the rate and mass distribution of objects impacting all three bodies has been about the same over the past 3.5 billion years (based on the age of the lunar lavas considered in Figure 6.11). Under this hypothesis, all three surfaces would be about the same age—on the order of 3.5 billion years or so. Thus the simplest explanation is that all three planets have experienced bombardment over about the same length of time by about the same population of objects.

The other interpretation is based on the presumption that the rate of impact cratering at the orbit of Mars must have been very much higher than at Mercury, since the objects that caused the cratering originated in the asteroid belt, just beyond the orbit of Mars. This interpretation requires that the ages of the smooth plains on all three planets be exactly adjusted so as to make the accumulated populations the same despite the different flux histories; the present similarity in accumulated impacts is deemed to be entirely fortuitous. Mercury's smooth plains

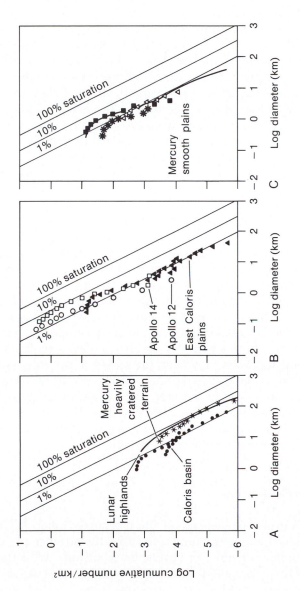

Figure 6.11

Moon and Mercury crater size-frequency distributions are expressed as the cumulative number of craters larger than a given diameter and compared with percentages of saturation. Equilibrium conditions (i.e., when the rate of crater production equals the rate of crater destruction) are attained for crater populations at 5 to 10 percent saturation. Key to Symbols: **(A)** Star = heavily cratered terrain; black dot = Caloris Basin; heavy line = lunar southern highlands. **(B)** Black triangles = plains east of Caloris Basin; open circles = Apollo 12 landing site; hollow square = Apollo 14 landing site. **(C)** Asterisk = plains-filling crater; hollow triangle = crater floor; black square = plains-filling crater; heavy line, plains outside Caloris Basin.

would be somewhat older than the Moon's and would have recorded the tail end of the heavier bombardment from earlier times; the age of the lunar maria are fixed at around 3.5 billion years, and the martian plains would be much younger, perhaps by a billion or so years.

If the first interpretation is correct and the smooth plains of Mercury, Mars, and the Moon are all really of comparable age (and nature has not by chance created the similarity in size and frequency of impact craters), then the objects striking the bodies must come from orbits whose aphelia are so distant from the Sun that the differences in distance to the orbits of Mars and Mercury are insignificant by comparison. Such orbits are typically those of comets. Indeed, the conclusion that the ages of the smooth plains on the three planets are comparable carries with it some implication that the objects have come from cometlike orbits. In this regard, the subject of asteroids versus comets as the source of the lunar impacts continues to be controversial and is still unresolved within the context of the lunar data alone.

There are some difficulties—some loose ends. On both the Moon and Mars, a finite interval of time is recorded by the different smooth surfaces, as shown by the difference in crater frequencies for different surfaces. On Mercury, no such clear differences in crater abundances have been seen, but superposition and overlap relationships show clear evidence of similar age differences. In this respect, Mercury's plains are like those on the Moon and Mars. However, the nature of the ridges of the smooth plains and the absence of regional patterns of extension outside the Caloris Basin indicate that the deformation of Mercury's plains has been quite different from that of either the Moon or Mars, hence the age and origin of the mercurian smooth plains remain subjects of debate.

Mercury's Senescence

Regardless of the precise age of the mercurian smooth plains, the planet as a whole is characterized by the absence of subsequent tectonic, volcanic, or atmospheric history, with the exception of the regional deformation of the Caloris Basin, which is clearly related to the formation of that structure. Of course, the presence of local tectonic or volcanic features on the unseen half of the planet cannot be excluded. But certainly atmospheric effects would be global, and therefore the possibility of an atmospheric history can be ruled out by existing data. Thus Mercury evidently became internally quiescent by the time the mercurian plains were formed, presumably about 3.5–4 billion years ago. In that regard, Mercury is conspicuously and profoundly different from Mars, Earth, and Venus, but instead bears a strong similarity to the Moon.

Thus we are presented with a paradox. If we look at the history of Mercury—the external history carved out on its surface as well as the past 3.5 billion years of its internal history—we find a strong resemblance to that of the Moon, despite the differences in internal constitution and orbital position of the two planets. This paradox and its ultimate solution is of major importance to our understanding of the entire family of planets.

FUTURE PROSPECTS

Overview

The most plausible history of Mercury's surface can be stated in terms of five stages. (1) Major chemical differentiation took place either during or very shortly after accretion. (2) During or after the decline of the original bombardment associated with accretion came a period of crater obliteration, perhaps through extensive volcanism. This either accompanied or was followed by the termination of late heavy bombardment. Global crustal contraction also took place at that time, perhaps due to core contraction. (3) Near the end of late heavy bombardment came the formation of the gigantic Caloris Basin, along with the creation of the characteristic basin-related mountains and sculptured terrain surrounding it. (4) Episodes of volcanic flooding, apparently during a fairly brief period, resulted in the filling of large areas, including the Caloris Basin and extensive areas east and north of it. (5) A long period of internal quiescence and

light cratering—spanning most of the planet's history—
followed the filling episodes. A cumulative impact flux comparable to that recorded on the lunar maria was recorded on the
mercurian surface during that stage.

Other interpretations of Mercury's history can be inferred
from the Mariner 10 data, although they require more *ad hoc*
assumptions. For example, the early crater-obliteration process
might have been catastrophic, or localized volcanic resurfacing
may have been prevalent. Thus late heavy bombardment need
not have been episodic. It is also possible that none of the
smooth plains are of volcanic origin; there is, after all, no specific morphological or chemical proof that they are.

The Nature and Origin of the Magnetic Field

Mercury proved to have a dipole magnetic field aligned with
its spin axis in a manner very similar to that of the Earth. However, Mercury's field has only about 1/100 the intensity (Fig.
2.5) of the Earth's field. The existence of the field, discovered
by Mariner 10, was unexpected. Mercury spins about its axis so
much more slowly than the Earth that the fluid dynamical
mechanism believed to generate the Earth's field did not seem
applicable to Mercury. The Earth's field is thought to arise from
electric currents associated with differential motions in its
spinning fluid metal core—that is, a "self-dynamo" effect.

There have been serious attempts to extrapolate the physical
model of the Earth's field to Mercury and show that indeed the
observed field could be produced on Mercury. Curiously, no
spacecraft have observed any planetary magnetic field on
Venus, even though it probably has a larger and hotter core
than Mercury. However, it does rotate even more slowly than
Mercury. Furthermore, if there are fluid motions within the
mercurian core capable of generating the magnetic field, the
core motions—or at least the associated heat flow—have not
produced any recognizable deformation of the planet's surface
for nearly 4 billion years. Finally, an active dynamo within
Mercury would very likely require a continuing source of heat
to keep it running. But because of the planet's small size, its
initial store of heat must have dissipated by now through radio-

ative cooling. Thus it is difficult to accept a simple Earthlike model for the mercurian field.

Nevertheless, proposed alternatives are even more difficult. For example, one possibility is that the present field is remanent—that is a fossil magnetic field left from an earlier time when the planet might have had an Earthlike dynamo. However, it seems highly probable that the temperature within most of Mercury's interior during that long time must have been above the Curie point (the temperature at which a substance loses its magnetism). Another alternative is that the mercurian field somehow has been induced as a result of Mercury's continued interaction with the solar wind. But it is difficult to reconcile this possibility with the fact that the present field exhibits symmetry about the rotation axis.

Perhaps the mercurian magnetic field arises from causes yet unimagined. Indeed, strong but unexplained remanent magnetizations have been observed in portions of the lunar crust. Perhaps we shall really have to gain a deeper understanding of how the Earth's own field is created to recognize how that mechanism can be reduced at the mercurian scale yet be missing from Venus. Wherever the truth lies, it is indeed fortunate that, among the inner planets, there is one whose magnetic field seems sufficiently similar to Earth's to allow useful comparison to be made as to origin.

The Next Step

Where next in the exploration of Mercury? How should mankind follow up these exciting clues about the nature of the planet and its relationship to the other inner planets? There seems to be little hope of further exploration from the Earth's surface except by more powerful radar. Conceivably, the equatorial surface features of Mercury conceivably could be mapped by radar over the coming decade, but only at low (10–100 kilometers) resolution. Experience with Mariner 9 on Mars has shown that this kind of resolution is inadequate to infer surface history. Furthermore, in order to understand Mercury's magnetic field fully, nothing short of continuing global measurements will do. Accordingly, the essential next step in the study of Mercury will be to place a spacecraft designed to survive in orbit around the planet for at least six months to a year.

Such a spacecraft would be able to complete the photographic mapping of the planet and to carry out specialized investigations to follow up such intriguing questions as the origin of the intercrater plains and the smooth plains. It is important to remember in this regard that Mariner 9, which followed three flybys of Mars, discovered totally unexpected surface features in the then-unexplored half of that planet. The Moon also provided a surprise when the crude camera of Luna III, in 1959, returned the first glimpse of the lunar farside, revealing that the dark maria are almost entirely restricted to the nearside hemisphere. Since Mercury has been observed over only the half of its surface that was illuminated during the Mariner 10 flybys, perhaps similar surprises will come with further exploration.

A principal objective of a mercurian orbiter should be to map carefully the magnetic field surrounding the planet and to continue that mapping throughout its orbit around the Sun. Mapping the field through a complete orbit is important because the (weak) magnetic field can be highly disturbed by the solar wind. The true detailed shape of the planetary field can be isolated only from measurements made over much of the mercurian orbit. In addition, more sophisticated measurements may be possible in which the electrodynamic response of the planet to fluctuations in the electric and magnetic fields of the interplanetary winds can be ascertained, thereby gaining valuable information on the electrical conductivity of the planet's interior. A definitive test for the presence of a fluid iron core may be possible in this manner.

Careful tracking of the satellite's orbit will provide extremely important information about variations in Mercury's surface gravity, and therefore variations in surface density. Ideed, it would be possible to search for mercurian equivalents to the mascons—the gravity anomalies associated with the circular lunar maria. Their presence and shape on Mercury might provide independent evidence of volcanic origin of the plains through analogy with the Moon; their absence could point toward a nonvolcanic origin of the plains.

If adequate payload proves to be available, it would be desirable to carry the full complement of instruments considered for a Lunar Polar Orbiter Mission. This set includes gamma and x-ray measurements from orbit that can map surface compositional variations. Indeed, by flying a set of instruments around Mercury similar to those that (by then) have been flown previously around the Moon, the power of Apollo sample analysis could be extended indirectly to Mercury. Comparative planetology of the inner planets would reach new sophistication under such circumstances.

SUGGESTED READING

Davies, M. E., S. E. Dwornik, D. E. Gault, and R. G. Strom. *Atlas of Mercury*. NASA SP-423. Washington, D.C.: U.S. Government Printing Office, 1978.

Dzurisin, D. "The Tectonic and Volcanic History of Mercury as Inferred from Studies of Scarps, Ridges, Troughs and Other Lineaments." *J. Geophys. Res.* **83**(B10):4883–4906, 1978.

Malin, M. C. "Observations of Intercrater Plains on Mercury." *Geophys. Res. Lett.* **3**(10):581–584, 1976.

Strom, R. G. "Mercury: A Post Mariner 10 Assessment." *Space Sci. Rev.* **24**(1):3–70, 1979.

Mariner 10 Preliminary Science Report. Science **185**:141–180, 1974.

Mariner 10 Imaging Science Final Report. J. Geophys. Res. **80**(17):2341–2514, 1975.

C. H. Simmonds, C. H. and R. B. Merrill (editors). Proceedings of the Conference on Comparisons of Mercury and the Moon. *Phys. Earth Planet. Int.* **15**(2/3):113–312, 1977.

Special Mercury Issue. Icarus **28**(4):429–609, 1976.

Note: The U.S. Geological Survey is currently publishing a series of geological maps of Mercury at a scale of 1:5 million as part of its Miscellaneous Investigations Map Series.

7

THE PUZZLE
THAT IS MARS

THE PUZZLE
THAT IS MARS

No other planet of the Solar System has held our fascination as much as Mars. In the eyes of the ancients, the Red Planet symbolized fire, battle, and power. It was not until the Renaissance that Mars was examined free of obvious fable. With the spread of the telescope in the early 1600s, scientists could see that planets were more than mere points of light in the sky. The first record of surface features on Mars was made by Christian Huygens, a Dutch physicist who, in 1659, sketched the martian feature later named Syrtis Major ("Great Quicksands").

By the mid-1800s, telescopic observations resulted in charts showing prominent surface markings. Some of these maps showed dark linear features. Giovanni Schiaparelli, director of the Milan Observatory, produced a detailed chart of Mars based on observations from 1877 to 1888. He named the dark areas after Egyptian gods, biblical lands, and places in the mythological underworld, or Hell. The brighter areas were named after deserts on Earth; many of these names are retained for albedo features on modern charts (Fig. 7.1).

Schiaparelli mapped more of the long, dark features than anyone preceding him. He named them canali—"channels" in Italian—and thus set in motion a controversy that persisted for nearly a century. "Channels" was mistranslated by later observers as "canals," a term that suggested artificial structures and conjured the possibility of intelligent life in the minds of many.

By the turn of the century, discussions and writings about Mars and "Martians" were nearly inseparable in both scientific and literary circles. Percival Lowell built the Lowell Observatory near Flagstaff, Arizona, and initiated observations of Mars that continue there today. Lowell's writings, particularly his book *Mars and Its Canals* (1906), fired the public imagination. The controversy over canals on Mars continued into the Space Age, in part stimulating early U.S. public support for exploration of the Red Planet.

Exploration of Mars beyond Earth-based studies began in 1965 with the successful flight of the U.S. Mariner 4 flyby, and continued with Mariner 6 and 7 flybys in 1969 and Mariner 9 Orbiter in 1971–1972. The Soviet flybys, Mars 1 (1965) and Zond 2 (1967), failed before arrival. The Soviet orbiters Mars 2 (1971) and 5 (1973) added modestly to the Mariner 9 results. The lander portions of the Soviet missions, Mars 2, 3, 4, and 6, were failures. The U.S. Viking 1 and 2 Landers and Orbiters (1976–1980) have culminated this extraordinary epoch of exploration with detailed surface analyses and observations as well as refined orbital observations.

These missions laid to rest the stories about *canali*. Most of the mapped "canals" simply did not exist; the remainder were either natural tectonic features or albedo patterns. The tremendous wealth of data returned by the Mariner missions, particularly the Mariner 9 Orbiter of 1971–1972, provided the basis for the first global geological studies of Mars; detailed geological mapping is in progress using high-resolution data from Viking Orbiter missions.

This chapter summarizes the history of the surface and atmosphere of Mars as it now appears, recognizing the changing and unpredictable nature of this fascinating subject.

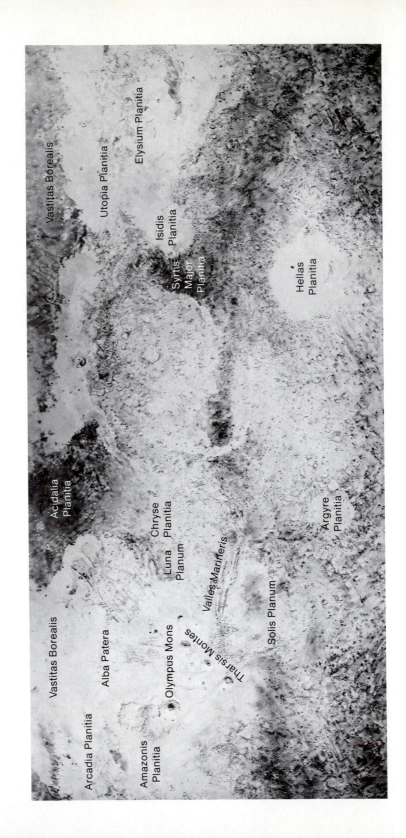

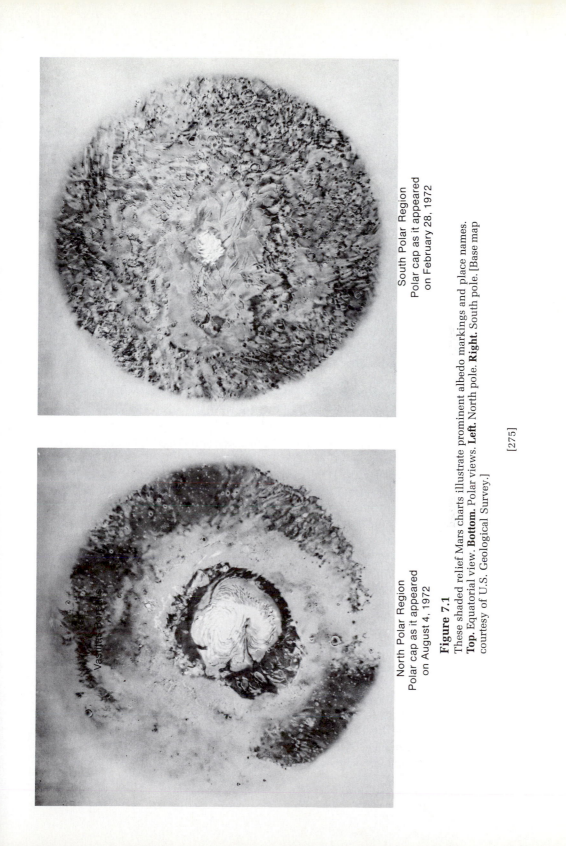

North Polar Region
Polar cap as it appeared
on August 4, 1972

South Polar Region
Polar cap as it appeared
on February 28, 1972

Figure 7.1
These shaded relief Mars charts illustrate prominent albedo markings and place names.
Top. Equatorial view. **Bottom.** Polar views. **Left.** North pole. **Right.** South pole. [Base map
courtesy of U.S. Geological Survey.]

[275]

THE OLDEST TERRAINS

Ancient Craters and Basins

Much of the southern hemisphere of Mars is made up of an assemblage of topographic features collectively designated *heavily cratered terrain* (Figs. 7.2, 7.3). This unit consists of clusters of large craters, multi-ringed impact basins, and large expanses of intercrater plains (Fig. 7.6). It appears to record an ancient bombardment history similar to that evidenced by the lunar highlands and the heavily cratered terrain of Mercury. Also present on the martian heavily crater terrain is evidence of contemporaneous redistribution of material, presumably by an early atmosphere.

Sixteen large impact structures (more than 250 kilometers in diameter) can be seen on Mars (Figs. 7.4, 7.5). Some are relatively fresh; others are extremely degraded. The rim deposits of these enormous features form arcuate ranges of mountains that show structural control and, in some instances, radial and concentric patterns. These are interpreted to be blocks of crustal material uplifted during the impact events.

The martian heavily cratered terrain does not exhibit as many craters as do the lunar highlands. On the average, there are 15 craters larger than 50 kilometers in diameter per million square kilometers on Mars, only a quarter the abundance recognized on the Moon.

The Intercrater Plains

The intercrater plains are characterized by low, rolling topography that is occasionally quite rough. Very small, fresh craters (many are probably secondaries) are abundant in some areas. Striking variations in crater morphology from region to region suggest corresponding regional differences in the properties of surface materials. Flow lobes and other features that resemble volcanic landforms suggest (but do not prove) that some inter-crater plains are lava flows (Fig. 7.6).

Stratification of the intercrater plains is suggested by the ter-raced eroded slopes within the Ganges Chasma of the Valles Marineris. A loosely consolidated material possibly rich in vol-atiles has been suggested for the host material. Elsewhere in the Valles Marineris, layers are seen within the wall formations (Fig. 7.7). A distinct dark unit seems to cap the sequence, which is consistent with the idea that the smooth surface con-sists of a later volcanic material that covers the older intercrater plains unit. In the walls of the canyons, beneath the sharp, darker brink, alternating bands of bright and dark material sug-gest layering, as do the differences in outcrop expression that can be seen downslope. Such layering has been observed in several widely separated locales extending over the entire length of the Marineris system, some 3000 kilometers.

Heavy Bombardment and Intense Atmosphere/Surface Interaction

The ancient martian environment recorded in the heavily cra-tered terrain must have been dramatically different from the one we see today. Early Mars was dominated not only by im-pact but probably by significant atmospheric erosion and vol-canism as well. The surface appears to have been continually modified during the period of heavy bombardment. Most of the large martian craters are clearly degraded (i.e., have "soft"-appearing rims; flat, featureless floors; no secondaries; and few continuous deposits). They do, however, display a range of morphologies and degradation that implies modification dur-ing an extended period of crater formation. The paucity of smaller craters in some areas of Mars, compared with the Moon and Mercury, suggests that ancient depositional or erosional processes buried or destroyed small craters and subdued large craters *before* the cessation of heavy bombardment.

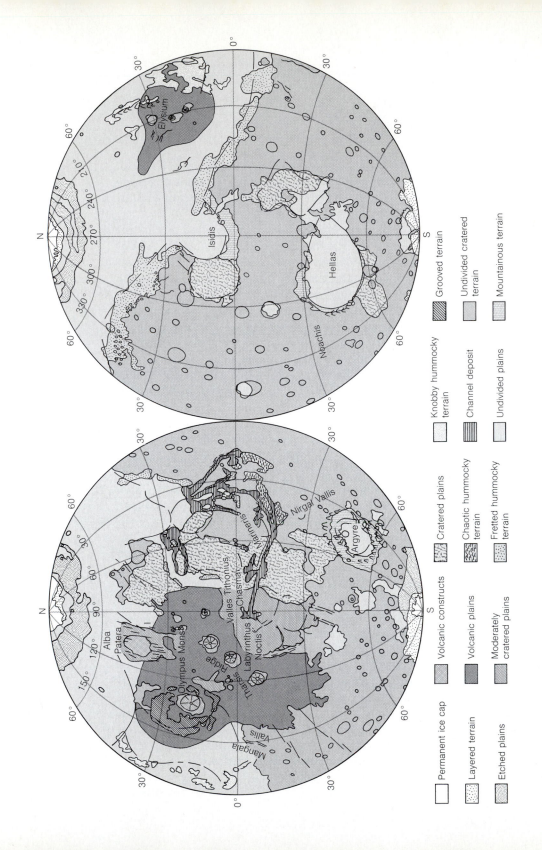

Permanent ice cap

Layered terrain

Etched plains

Volcanic constructs

Volcanic plains

Moderately cratered plains

Cratered plains

Chaotic hummocky terrain

Fretted hummocky terrain

Knobby hummocky terrain

Channel deposit

Undivided plains

Grooved terrain

Undivided cratered terrain

Mountainous terrain

Alba Patera

Olympus Mons

Tharsis Montes

Noctis Labyrinthus

Valles Marineris

Tithonius Chasma

Mangala Vallis

Nirgal Vallis

Argyre

Elysium

Isidis

Hellas

Noachis

Figure 7.2

Geological features of Mars are depicted as they were inferred from Mariner 9 pictures. The northern hemisphere of Mars has many young volcanic structures; the crust of the southern hemisphere is older and heavily cratered. The permanent ice caps are composed chiefly of water ice that rests on top of layered terrain; the seasonal ice caps are composed of carbon dioxide ice. The layered terrain is a relatively young collection of successive blankets of dust and ice deposited near the poles. The etched plains are irregular pitted deposits that have been eroded by the wind. The volcanic structures are largely shields, domes, and volcanic plains with few craters. The cratered plains are the most densely cratered of the plains regions on Mars. Chaotic hummocky terrain is a depressed area consisting of disrupted and tilted blocks of the martian crust. Fretted hummocky terrain is a lowland region with numerous isolated mesas bordered by cliffs. Knobby hummocky terrain is an isolated region of knobs, each about 10 kilometers across. The channel deposits are the smooth floors of channels and canyons, probably composed of material deposited by water, wind, or landslides. The undivided plains have a generally scoured appearance, full of irregular ridges, scarps and channels, and are sparsely to moderately cratered. Grooved terrain is a line of mountains next to Olympus Mons that range between 1 kilometer and 5 kilometers wide and form an arc about 100 kilometers long. Undivided cratered terrain is a region of densely to moderately cratered uplands and is the most ancient part of the exposed surface of Mars. The mountainous terrain is the rugged area adjacent to an ancient impact basin [From "Mars," by James B. Pollack [after maps by Thomas A. Mutch]. Copyright © 1975 by Scientific American, Inc. All rights reserved.]

[279]

A

Figure 7.3
(A) This Mariner 6 wide-angle photograph displays a large expanse of cratered terrain in Deucalionis Regio. The large crater, Flaugergues, is located near 17°S latitude and 340°W longitude and is about 200 kilometers in diameter. Note the essentially smooth, featureless textures of the intercrater surface at this scale. **(B)** This Viking Orbiter 1 mosaic shows at higher resolution the crater Flaugergues and part of the surrounding terrain seen above. The intercrater plains are seen here to be much more rugged and dissected than was apparent from the earlier photography. Note also that the floor of the crater has been partly filled with lunar mare-like plains.

B

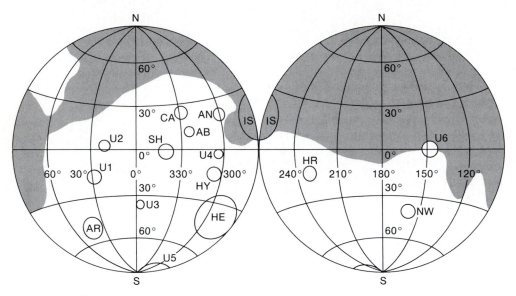

Figure 7.4
Martian craters and multiringed basins greater than 250 kilometers in diameter are plotted on this Lambert equal-area map projection. Similar craters in shaded regions have been destroyed or hidden during plains formation. Circles indicate approximate size only. AR = Argyre; SH = Schiaparelli; CA = Cassini; AB = Arabia; HY = Huygens; AN = Antoniadi; HE = Hellas; IS = Isidis; HR = Herschel; NW = Newton; U1–U6, unnamed.

Figure 7.5
In this oblique view of Mars looking southeasterly toward the horizon, some 19,000 kilometers away, Argyre Planitia is the relatively smooth plain at top center, surrounded by heavily cratered terrain. Argyre is a large, ancient impact basin that lies 50° south of the equator and has long been known to telescope observers as a region of occasional clouds. On this day, the martian atmosphere was unusually clear, and craters can be seen nearly to the horizon. However, the brightness of the horizon is due mainly to a thin haze. Above the horizon are detached layers of haze 25 to 40 kilometers high, thought to be crystals of carbon dioxide (dry ice).

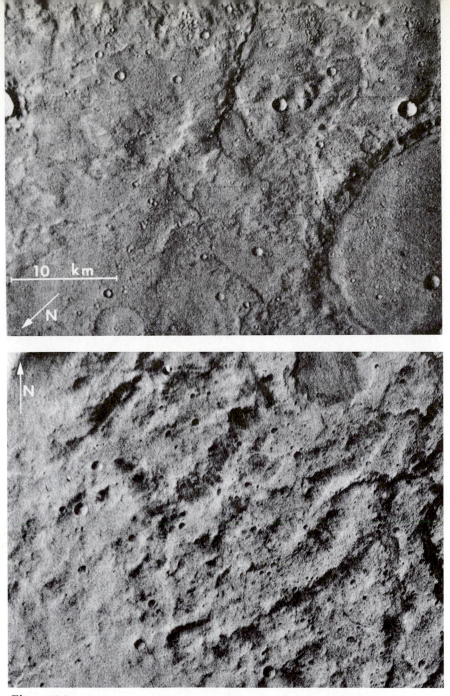

Figure 7.6
These high-resolution Mariner 9 photographs show some of the complex nature of the morphology of the intercrater plains in the martian equatorial regions. Escarpments, ridges, isolated hills, channel-like forms, and wavelike hummocks combine with craters and crater deposits to form a confusing pattern of topographic forms.

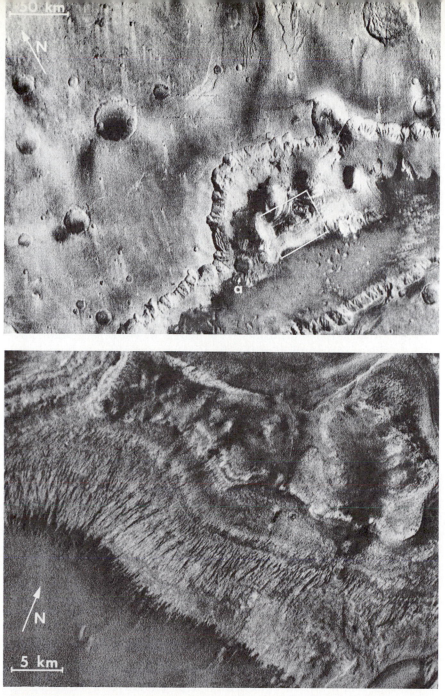

Figure 7.7
Top. This Mariner 9 A-frame shows a layered deposit within Ganges Chasma (6°S, 50°W) and outlines the location of the accompanying high-resolution frame. Note the craterlike form (a) west of the layered mesa. **Bottom.** In this high-resolution B-frame the intratrough mesa is about 2 kilometers high, and its summit is essentially level with the plain in which the trough formed.

These morphological aspects of the ancient cratered terrains on Mars indicate that the martian atmosphere was once appreciably denser than that presently observed—dense enough to be capable of transporting considerable sedimentary material. Viking isotopic analyses of atmospheric gases are consistent with these indications of a denser ancient atmosphere on Mars. Impact-generated debris (analogous to the lunar surface materials), weathering products, and volcanic ash are all possible constituent materials to have been transported and deposited by an early dense atmosphere. Layering, like that indicated in the walls of Valles Marineris, may imply episodic environmental variation—atmospheric, volcanic, or both. Since the extensive redistribution of this fine material appears to have terminated with the end of the heavy bombardment, Mars evidently lost its early "dense" atmosphere at that time, through escape of atmospheric gases from Mars' gravity field or through recombination with surface materials. The seeming association of the demise of the dense atmosphere with the cessation of heavy bombardment could mean that the atmospheric volatiles were being steadily released during the bombardment process, either from the impacting (cometary) objects themselves or from within the surface layers of Mars as a result of impact. Alternatively, the planetary heating from such impacts conceivably may have triggered the release of volatiles from the bulk of the planet. In any case, such volatiles apparently remained only briefly as part of the martian atmosphere; gases that are cosmically abundant but unstable (on Mars), such as hydrogen and ammonia, conceivably might have been among those lost.

Formation of Fretted and Chaotic Terrain and the Channels

The end of the heavy bombardment of the martian surface marked a period of transition from an active to a quieter erosional environment in which some of the most "unearthly" terrains on Mars were formed. The results of three processes—fretting, chaotic terrain formation, and channel formation—will be examined for clues to this transition period of martian history.

Fretted terrain occurs in localized regions along the margins between cratered "uplands" and uncratered "lowlands" (Fig. 7.2). It is characterized by abrupt and planimetrically complex escarpments with numerous islandlike outliers that resemble mesas and buttes. Originally described from Mariner 9 pictures, the low areas between the outliers appeared to be relatively smooth and featureless; high-resolution Viking images, however, show subparallel ridges and grooves in many areas, suggesting creep of near-surface materials such as might result from freeze/thaw of interstitial ice. Structural control of the erosional processes seems likely in view of prevalent trends displayed by steep-walled escarpments and channel-like embayments. Fretted terrain is thought to form by scarp recession attributed to undermining by evaporation of ground ice or ground water (i.e., "sapping"). Debris shed from the escarpments could have been removed by eolian deflation, provided the constituent materials are of intrinsically small particle size or uniformly reduced to small sizes by weathering.

Chaotic terrain is confined primarily to a region within $\pm 15°$ of the equator and between 0° and 60°W longitude, an area informally called the "Chryse Trough" because of its topographic form (a broad depression trending roughly north-south). Chaotic terrain consists of regions with a haphazard jumble of large angular blocks and arc-shaped slump blocks on the steep, bounding escarpments (Fig. 7.8). Some areas of chaotic terrain are irregular in plan view and others circular and crater-like; some areas are contained within craters. Chaotic terrain has been attributed to localized collapse by the removal of subsurface material (either magma or ground ice). Extensive subsequent modification of collapsed landforms suggests that weathering and transportation processes similar to those proposed for "fretting" were active as well. However, the removal of material may have been less complete or effective than in fretted regions, as suggested by the gradation from smooth channel floors to the rough, jumbled, positive floor topography of the chaotic terrain. The restricted vertical development of chaotic terrain suggests that the erosional processes were effective only to a limited depth, perhaps associated with the state of interstitial water in the subsurface layers.

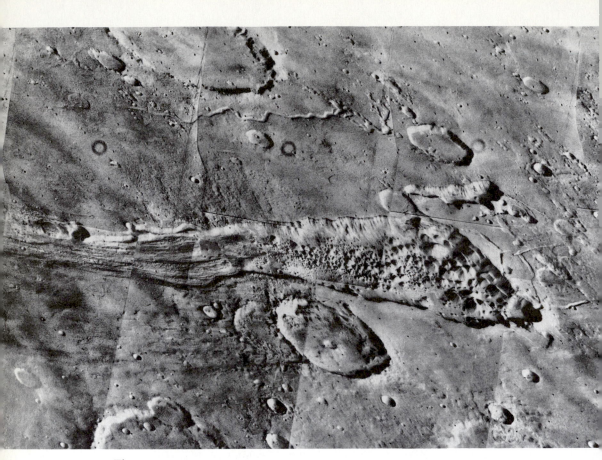

Figure 7.8
The head of an outflow channel is seen in this oblique southward view of Capri Chasma. The headward region consists of chaotic terrain that appears to have developed by the melting of subsurface ice and the subsequent release of floods to form the channel to the left. The area shown is about 300 kilometers by 300 kilometers.

Channels on Mars have a wide variety of forms and configurations (see Chapter 3). They seem to have developed in three epochs. The oldest are small gullies and dendritic networks confined to the old heavily cratered terrain (Fig. 7.9). The patterns of branching and of the junction angles of their tributaries are similar to certain kinds of drainage patterns on Earth. The lack of "pirated" streams, the preservation of old crater features adjacent to the channels, and the lack of extensively modified terrain between the channels suggest that the channels were active for a relatively short period of time. Impact craters superposed on the channels show that these features were formed

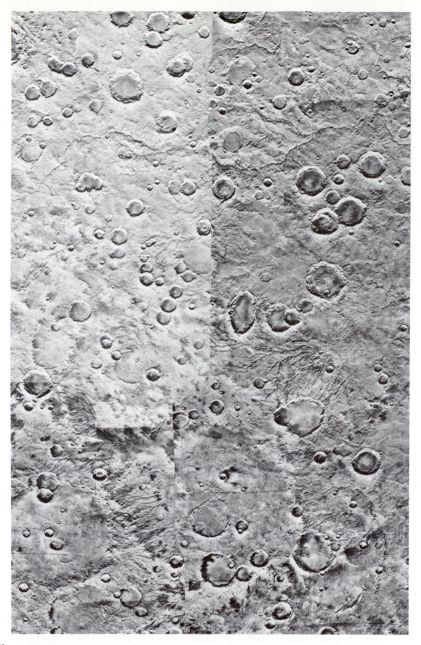

Figure 7.9
This Viking Orbiter mosaic of part of the southern hemisphere of Mars depicts impact craters, intercrater plains, and a network of dendritic channels. These and other channels in the region constitute the oldest channels on Mars and are considered to be fluvially eroded features formed billions of years ago during an episode in Mars' history when the atmospheric pressure and temperature were high enough to permit liquid water on the surface. Area shown is about 1200 kilometers by 800 kilometers.

early in the history of Mars; they appear to be absent from all younger terrains. Hence they are best interpreted as having been cut into the heavily cratered terrain either contemporaneously with its formation or immediately following the cessation of heavy bombardment. The morphology and planimetric outline of these gullies is most similar to that formed on Earth by sapping erosion associated with emanation of water from an underground water or ice table.

A second characteristic group of channels are the enormous "flood" features (Figs. 7.10, 3.28, 3.29, 3.30) first seen by Mariner 9. Nearly all of these features originate in chaotic terrain and terminate in the Chryse Basin. These features may have been completely filled by rapidly flowing water at their sources during their active formation, as evidenced by the fact that they emerge full size from their source regions. They have few, if any, tributaries, and they have scour marks on their upper walls and on the adjacent terrain. Moreover, teardrop-shaped islands with possible vortex-scours—features that form only in fluvial channels—are exhibited. The required discharge rates for such channels seemingly would be in the range 10^7–10^9 meter^3/sec, equal to or exceeding the discharge of the Ice Age Missoula Flood and far exceeding any river flow currently on Earth.

The youngest category of martian channels are apparently valleys that were probably initially cut by channels but which seem to reflect more prolonged subaerial erosional activity, including some eolian erosion.

End of Heavy Bombardment and Diminution of Atmosphere/Surface Interactions

Recent investigations suggest that all three types of landforms— fretted terrain, chaotic terrain, and at least some channels— are among the oldest on Mars. Few of these features formed during later epochs, which indicates that the intensity of erosional activity diminished abruptly very early in the planet's history, presumably in association with a transition to a more rarified atmosphere less suited to the accumulation of surface water.

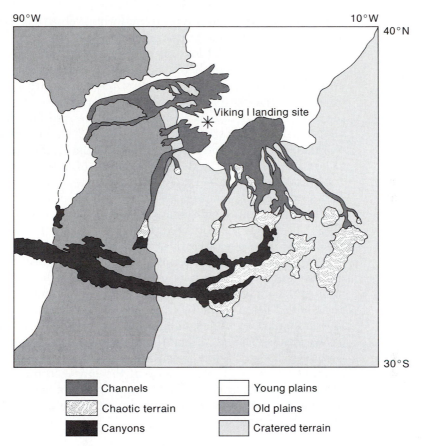

Figure 7.10
The geology of the region around the Chryse Basin, where most of the large flood features of Mars occur, is shown in this schematic. The channels start mostly in areas of chaotic terrain and extend downslope into the Chryse Basin, where they ultimately merge with the lava plains. The asterisk indicates the site of the Viking Lander 1. [Courtesy of Michael H. Carr, 1980.]

The planimetric and topographic form of chaotic and fretted terrains and of the flood channels suggests that some or all of the lithospheric volatiles (presumably water) necessary for their formation was probably supplied by the intercrater plains unit. These landscapes may reflect the temporary release of previously buried (or incorporated) atmospheric volatiles or the initial release of new volatiles from the interior. There are no comparable terrestrial analogs to this inferred circumstance on Mars. Surface runoff from rainfall and the sudden discharge of ponded water are the dominant causes of flooding on Earth.

In summary, Mars had a complex and unusual early history. The stratigraphic relations of plains units, cratered terrains, and channels imply significant atmospheric and fluid erosion (Fig. 7.11). That the source and disposition of the once-large quantities of surface water or other volatiles is unknown constitutes a principal question about Mars. The next oldest terrains—old volcanic plains that overlap the oldest heavily cratered terrains and which exhibit impact crater abundances suggesting an age of 3 billion years (by lunar standards)—exhibit no direct morphological evidence of having experienced modification by an atmosphere significantly different from that of today. However, there is some evidence supporting the notion that channel cutting did affect some of the old volcanic plains in certain places. In any case, the transition of the martian atmosphere at the end of heavy bombardment was a principal event in the planet's history.

VOLCANIC PLAINS AND TECTONISM

Following the cessation of heavy impact cratering and the formation of the intercrater plains, extensive volcanic plains were emplaced in several regions of Mars. The central volcanoes of the Tharsis region appear to be built on a base of earlier volcanic plains; similar relationships are seen in the Elysium region. In addition, similar plains units occur around Alba Patera

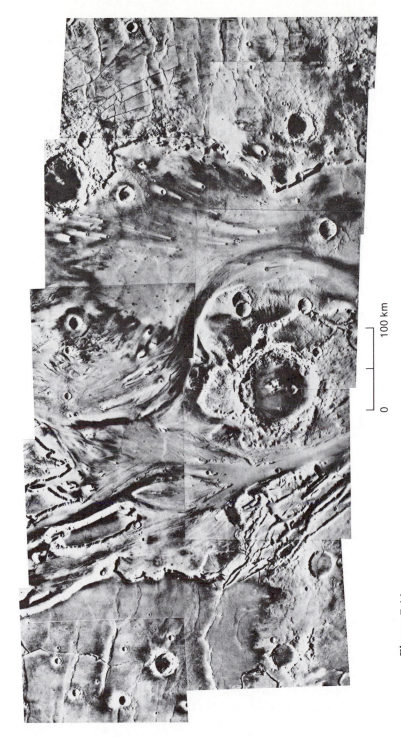

Figure 7.11
Viking Orbiter 1 mosaic of the region between Lunae Planum and Chryse Planitia includes examples of most of the major geological processes that have shaped the surface of Mars. Several large craters (the largest, in the lower center, is about 100 kilometers across) represent impact cratering; mesa-like structures on the west (left) side are remnants of flood lavas, representing volcanism of "intermediate" age in the history of Mars; fractures on the east (right) side represent crustal extension, or tectonism. Surface modification is represented by the enormous outflow channels in the middle of the mosaic, and the light crater streaks represent active wind processes. At least two channeling events are evident in this mosaic; note that the ejecta on the east side of the large impact crater is superimposed on the channel floor to the right, yet, on the west side, the ejecta is eroded by the left-hand channel. Thus the crater impact evidently occurred in the time between the formation of the east channel and the west channel.

0   100 km

and in the region of Lunae Planum and Chryse Planitia (Fig. 7.12). These units are characterized by a paucity of large (more than 20 kilometers) craters in comparison with the older cratered surfaces, which is indicative of their "intermediate" age.

Mars experienced a period of intense fracturing contemporaneous with or slightly after the formation of these volcanic plains. Faulting appears to have been extensional (i.e., displacement is vertical and at angles steeper than 45°). Neither significant strike-slip nor thrust faulting has been recognized on Mars. The proximity of the fracture systems to the enormous topographic rises (on which the Tharsis and Elysium volcanics are situated) suggests a genetic relationship between tectonic and volcanic activity. The majority of the martian tectonic features are arranged radially to the dominant Syria Rise (which includes the Tharsis region).

In some places, the development of tectonic features seems to have been highly localized, with surfaces intensely fractured in one area and smooth and unbroken only a few kilometers away in the same unit. The absence of fractures may not always imply superposition by younger deposits, but the alternative possibility, very highly localized stresses, is not easily explained by terrestrial analogy.

When Did the Tectonism Occur?

The tectonic history of Mars, especially in the Tharsis region, must be linked to the internal history of the planet. The key question here concerns the timing of tectonic activity in relation to the volcanic history.

The Tharsis region displays numerous fracture systems that fan out from the main topographic rise. Some fractures cut extremely old cratered terrain, whereas nearby old volcanic plains are undisturbed. Similarly, some volcanic plains of intermediate age are fractured, whereas adjacent plains—only slightly younger—appear pristine.

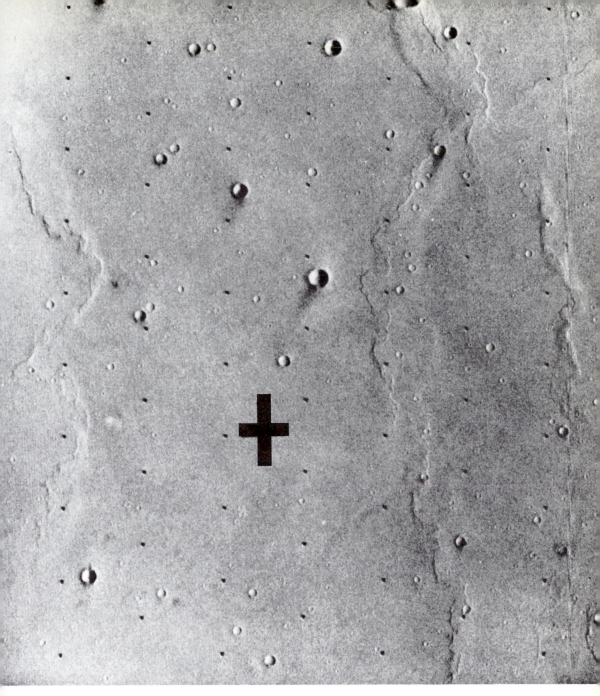

Figure 7.12
Photograph taken July 9, 1976, by the Mars-orbiting Viking 1 shows an area in the western part of Chryse Planitia. Mare ridges and impact craters in the lava flows that cover the area are prominent. Similar ridges are common in the mare surfaces of the moon. Cross marks the approximate position of the Viking Lander 1 site.

Three alternative implications of these observations are that (1) deformation was highly localized, (2) widespread fracturing occurred but is preserved in some units but not in others, and (3) tectonic activity in Tharsis was cyclic or episodic. However, the similarity in surface morphologies of the fractured plains and the adjacent unfractured plains argues against the notion that they may consist of materials of radically different internal properties. Extremely localized deformation is intuitively unattractive but cannot be ruled out completely. A likely explanation is that tectonic activity occurred periodically throughout martian history and that the present distribution of fractured surfaces reflects the time scales both of deformation and of formation of new surfaces by deposition and/or erosion.

Origin and Evolution of the Chryse Basin

Gravity and topographic data suggest to some that the depressed regions of Amazonis and Chryse are genetically and temporally related to the Syria Rise. Particularly within the Chryse Trough, an arcuate depression 2000 kilometers in length, evidence may point to considerable age. For example, all the major channels within the Chryse region are aligned with the present topographic gradient. Thus the regional topography must predate the channels. Most of the channels are relatively old features. Hence the Chryse Trough is old, and, by the suggested relationship, so also must be the Syria Rise.

Other data hint that additional phenomena also may have shaped the Chryse Trough. Although this depression appears to be a continuous feature, a major portion of its southern section consists of what may be a large but severely degraded impact basin. The arrangement of channels around the basin suggests a closed drainage system. The northern part of the trough is marked by the semicircular margin of the Chryse Basin. Thus, gravitational and topographic analyses notwithstanding, the origin of the Chryse Trough remains uncertain, although it is probably of great antiquity.

MARTIAN SHIELD VOLCANOES

Few geological processes stir so much excitement as erupting volcanoes. To stand on the rim of a caldera and see churning, molten rock, born far below the surface of the Earth, stimulates both primitive fear and rapt fascination. For those who have experienced active volcanism, the pictures of Olympus Mons on Mars (Fig. 4.30) are truly overwhelming. Despite the probability that the martian volcanoes seen in photographs have been extinct for millions of years, those pictures trigger images of enormous, boiling, turbulent calderas spilling lava over the flanks of these largest known volcanoes.

The large, central-vent, constructional volcanoes of Mars (Fig. 7.2) occur in the general regions of Tharsis and Elysium; other possible central-vent volcanoes, called patera, occur as individual structures in isolated regions of the planet. Although the origin of the patera is uncertain, their morphology suggests volcanism. At least some patera are probably very old, eroded or modified shield volcanoes.

The Tharsis Shields

The greatest concentration of shield volcanoes is in the Tharsis region, dominated by three great constructs—Arsia Mons, Pavonis Mons, and Ascreaus Mons—aligned northeast-southwest along the summit of a ridge 2000 kilometers long called the Tharsis uplift (Fig. 4.14). The ridge stands some 9 kilometers above the average martian base level; each of these volcanoes rises nearly 10 kilometers above the ridge, making them and Olympus Mons the highest places on the planet. During the great dust storm of 1971, the summits of the shields were discernible because they stood above the dust clouds. Olympus Mons lies about 1000 kilometers northwest of the Tharsis ridge; its 600-kilometer base diameter and 25-kilometer height make it the largest shield volcano on Mars, much larger than any shield on Earth; in the entire Solar System, it may be rivaled only by the recently discovered active volcanic features on Jupiter's moon, Io.

The great height and mass of the Tharsis shield volcanoes provide clues to the nature of the interior of Mars. A large hydrostatic pressure was required at depth to push magma to the summits of these shields. By terrestrial analogy, the magma chamber must have been located beneath a thick lithosphere, about 200 kilometers thick, roughly twice that of the Earth. However, strong mantle upwelling on Mars could be partly responsible, in which case the lithosphere in the Tharsis region might be correspondingly thinner.

From crater-count data and the general morphology of the Tharsis shields, especially as revealed in Viking photography, the sequences of eruption and the construction of the shields appear to be quite complex. Fresh-appearing lava flows are identified with all the Tharsis shields. Cross-cutting relations and superposition show an elaborate sequence of eruption and tectonism.

Moreover, many of the plains surrounding the shields are postulated to be relatively young lava flows related in origin to the shields, rather than being of separate origin. A varied and extended volcanic history is apparent.

Other Martian Volcanoes

The gentle (6°) slopes on the flanks of the shields, the morphology of the individual flows, and the presence of lava tubes on the Tharsis shields all point toward rather fluid lavas, probably basaltic in composition. However, besides the four large shields, there are several smaller shields and dome-type volcanoes in the Tharsis region (Fig. 7.13). The domes are particularly intriguing, as they may represent either a different style of eruption or lavas of a different composition from those that produced the shields The slopes of the domes are greater than those of the shields and appear to be somewhat convex. This morphology suggests construction by more viscous lavas, perhaps blocky flows. This morphology could also result from

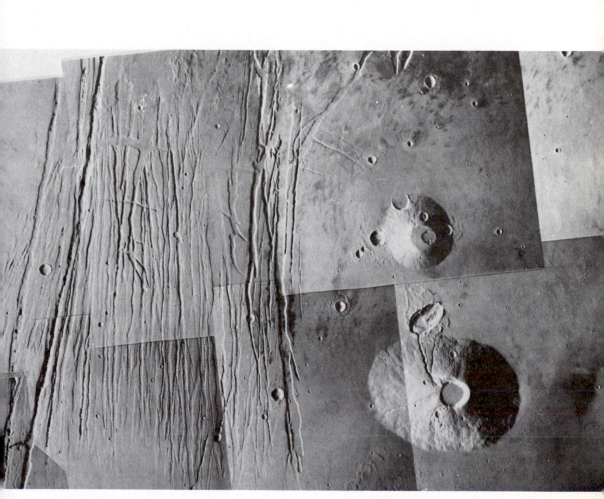

Figure 7.13

Ceranius Tholus (lower right) and Uranius Tholus (upper right) are two shield volcanoes of the northern Tharsis region that have been flooded by younger plains lava flows. The elongate crater between the two volcanoes of uncertain origin is younger than both the plains and the volcanic shield. To the west (left) of the volcanoes, the plains have been severely fractured, forming the Ceranius Fossae; this fracture set is part of the prominent north-south system that extends from Tharsis to Alba Patera. Area shown in this Viking Orbital mosaic is about 550 kilometers across.

short flows, reflecting lower rates of effusion than those that produced the shield volcanoes.

The other principal area with prominent volcanoes is the Elysium region (Figs. 4.35, 7.2). Elysium Mons—15 kilometers high and 200 by 300 kilometers across—and Hecates Tholus both have complex summit calderas, radial rift zones, and features that appear to be collapsed lava tubes and channels. Elysium Mons has steeper slopes than the Tharsis shields, and the distinctive profile of Elysium Mons more closely resembles an "evolved" shield volcano on Earth than does any other martian feature. The Elysium volcanoes have been compared morphologically to the volcanoes of the Tibesti Plateau of north central Africa.

The Patera

Some patera of Mars are the most puzzling of the probable volcanic features. The term "patera" (Latin for "saucer") was coined from the Mariner 9 observations of depressions with scalloped walls and a sometimes striking starburst pattern of ridges and channels. Although most patera appear to be eroded shield volcanoes (Fig. 7.14) or shields that have been partly buried, Alba Patera appears to be unique to Mars. Alba is more than 1600 kilometers across and has a central, complex caldera-like depression. The morphology and association with other volcanic structures suggest that it is volcanic in origin. However, there is nothing analogous to this style of eruption on Earth, the Moon, or Mercury.

During the search for a landing site for the second Viking Lander, Viking Orbiter 2 photographed the western flank of Alba Patera and provided the opportunity to study the feature in some detail. The flanks of Alba are made up of extremely complex intertwining lava flows, some closely resembling the sheetlike Imbrium lava flows on the Moon. Other flow textures on Alba include complex terraces made up of individual fingers of lava several hundred kilometers long. They appear to have been fed in a manner similar to that observed in basaltic volcanism in Hawaii. The scale of the martian features is orders

Figure 7.14
The volcano Apollinaris Patera, at 8°S, 186°W, is probably quite old, as is suggested by the number of superimposed impact craters and the degraded character of the flanks of the volcano. As such, it represents early-stage volcanism on Mars. Width of area shown in the Viking Orbital mosaic is about 400 kilometers; north is toward upper left.

of magnitude larger, although the morphology is essentially the same.

Still another type of lava flow on Alba Patera consists of dozens of lava tubes and tube-fed flows. Similar to those on Olympus Mons, the tube-fed flows on Alba Patera are recognized by an axial ridge along the center of the flow, frequently marked by collapse of the lava tube roof. Some of these flows can be traced for more than 400 kilometers. A rough estimate of the total volume for a *single* average tube-fed flow shows that it would have taken at least several thousands of years for the same volume of lava to be emplaced at eruption rates typical here on Earth. Yet Alba is made up of many such flows of similar age. This fact poses interesting problems. Either the eruption rates were much higher than those that produce similar structures on Earth or the duration of eruption for single flows and flow units on Mars—and Alba Patera in particular—may have been much more prolonged than for any volcanic features on Earth.

Relative ages for the patera, as determined from their relation to other martian features, indicate that some formed very early in the planet's history. For example, Hadriaca Patera and Amphitrites Patera are found on the "rim" of the Hellas Basin and appear to have formed shortly after the formation of that ancient basin. Others, such as Biblis in the Tharsis region, appear to be younger.

The relative ages of the volcanic constructs of Mars span a considerable portion of the planet's geological history. Hence internal melting and extrusion have transpired over perhaps billions of years. Questions to be pursued include how the interior could stay volcanically active for so long and what gaseous contribution to the atmosphere has resulted from this long and extensive release of lava to the surface.

Internal Heat and Mantle Convection

As with all the Earthlike planets, the earliest history of Mars—from accretion to the time when surface features were preserved—is very poorly understood. We must rely on theoretical calculations and considerations of the planetary mass, density, and other geophysical characteristics to derive models for this period. From such considerations, it appears that the time scale for planetary accretion was short and therefore rapid enough for gravitational energy in the form of heat to be stored within the accreting body.

A second stage of global internal activity may have occurred (according to some models) a billion years or so after the first stage, or about 3.5 billion years ago, when melting is inferred to have occurred in the upper mantle. It may have been during this stage that early volcanism took place to form flood-type flows, and perhaps also to produce early shield volcanoes seen today as some of the patera. Alternatively, Mars may have undergone only gradual cooling since formation.

Examination of the geodetic shape of Mars reveals a pronounced bulge, the Syria Rise, centered at 8°S 105°W; it stands some 7 kilometers higher than the average elevation for the planet. The Tharsis Ridge makes up part of this rise. Gravity studies of the bulge, however, show that the elevated area is not compensated by an underlying "root" of low-density rock, as are almost all mountain ranges on Earth. The Syria Rise on Mars exhibits *excess* gravity over that expected if density-height compensation operates as it does on Earth (isostasy). Either a very thick lithosphere is required, or some dynamic mechanism must hold the mass at a high level. One such dynamic mechanism could be *mantle convection,* in which heated (and therefore less dense) upwelling cells converge beneath the crust, push it upward, and support it at a topographically high level. Another possibility is that of *chemical diapirism,* in which a plug of less dense material, differentiated deep in the mantle, pushes up on the overlying material as it attempts to migrate to the surface.

There is no direct evidence to distinguish whether the excess gravity is the result of a thick lithosphere or either of the two dynamic mechanisms, but several lines of indirect evidence suggest that deep-seated uplift is involved. The center of the bulge coincides with the Noctis Labyrinth (Fig. 7.2), an immense area crisscrossed with grabens and related fractures. These tectonic features result from tensional forces in the crust— just the type expected from updoming. Moreover, this is the area where the Tharsis shields are found—volcanoes of probable basaltic composition. By analogy with Earth, regions of upwelling, such as the Mid-Atlantic Ridge and the postulated "plume" beneath the Hawaiian Rise, typically are sites of basaltic volcanism. Furthermore, the form of the Tharsis shields is consistent with extrusion of an enormous volume of lava of unchanging chemical composition over as much as a billion years.

In a less pronounced manner, the Elysium area's tectonic and volcanic patterns are similar to those of the Syria Rise and the

Tharsis volcanoes. The Elysium area "bulges" about 4–6 kilometers above the planetary average elevation and displays tensional fractures and grabens, although to a lesser extent than the Tharsis uplift.

Mantle convection is believed to be the driving force for plate motion on Earth. If mantle convection exists on Mars, is there also plate-type tectonism? In the sense in which that term applies to the Earth, the answer is no. Mars does not show colliding plates, subduction zones, or extensive belts of folded sedimentary rocks. Presumably a thicker and therefore stronger lithosphere on Mars (200 kilometers versus about 60 kilometers on Earth) has retained its structural integrity despite the stresses induced from mantle convection.

Valles Marineris—Crustal Extension on a Grand Scale

The canyon systems of Mars were one of the many surprises first revealed by Mariner 9. Stretching more than 4000 kilometers along the equator, the main canyon system displays a dominant east-west structural imprint that includes grabens, strings of craters, and a prominent ridge that runs down the floor of the canyon (Fig. 7.1). More than 4 kilometers deep in many places, and ranging 80–100 kilometers in width, the Valles Marineris are probably the result of large-scale block faulting and modification by various agents of erosion. The age of the canyon system, estimated on the basis of crater counts, appears to be similar to that of the Tharsis volcanoes. The origin of the two landforms could be linked: withdrawal of magma from beneath the region in which the canyons formed may have supplied the Tharsis volcanic uplift.

Although the detailed lithology of the canyon walls cannot be known from remote observations alone, some of the cratered terrain adjacent to the canyonlands appears to be mantled, perhaps with eolian sediment. Layering and differences in outcrop form suggest that parts of the canyon walls may be composed of sedimentary rock or, possibly, wide sheets of volcanic deposits. A sedimentary sequence conceivably could hold substantial quantities of water, perhaps stored in a manner similar to that of terrestrial ground ice. It has been suggested that following a period of initial east-west faulting—possibly associated with the uplift of the Syria Rise—the canyon was enlarged by a complex array of erosional processes. If ground ice were exposed to the atmosphere, its water would sublime, making the sediment a loose, friable material subject to mass wasting and eolian removal. Even today the canyons are sites of dust activity, evidenced by eolian streaks and dune fields on the floor.

Volcanism and the Evolution of the Martian Atmosphere

The history of the atmosphere of Mars is intimately linked with the history of the surface and interior. Planetary atmospheres may be produced during (1) initial accretion of the body; (2) subsequent bombardment by volatile-rich objects; (3) early differentiation, when gases trapped during accretion could be released; and (4) later volcanic eruptions, when gases could be "leaked" from the planetary interior. A big question about Mars has to do with the quantity of volatiles that were released in association with the extensive and enduring volcanic eruptions.

Unfortunately, even on Earth the amount of gases released by volcanic activity is not known. Most volatiles associated with present-day terrestrial eruptions are recycled from crustal rocks and the atmosphere. However, even if the volcanic emissions over geological time contained only 1 percent of genuine juvenile volatiles, that would account for all the volatiles in the Earth's atmosphere as well as those held in the hydrosphere and rocks.

The style of volcanism in the Tharsis and Elysium regions, which may have prevailed throughout much of martian geological history, probably consisted of sporadic episodes of eruption. During these periods, magmatic volatiles may have been released that could have added significantly to the total martian atmosphere. However, most of the channels *pre-date* the Tharsis and Elysium volcanics and, possibly, the volcanism of the intercrater plains and patera as well. Hence the early, dense atmosphere inferred from the fluvial character of channels probably cannot be attributed to the outgassing associated with the volcanic extrusions that now so characterize Mars' surface, as that activity took place too late in the planet's history.

Viking lander analyses suggest a complex history of outgassing on Mars. To some scientists, comparisons of the ratios of the noble gas isotopes Argon 40 and 36 on Mars and on Earth indicate that outgassing on Mars took place at least hundreds of millions of years after the planet formed. In addition, the ratio of nitrogen isotopes suggests that a significant amount of that gas has somehow escaped. By analogy with abundances of terrestrial volatiles a denser early atmosphere is indicated. How this atmosphere evolved with time is a subject of great interest, especially in light of the geomorphic evidence that suggests an early dense atmosphere.

POLAR PHENOMENA AND THE YOUNGER MARTIAN DEPOSITS

Landforms in the polar regions reflect both their erosional history and depositional origin. The polar units are of two main types, layered and unlayered. It has been proposed that the layered units are eolian deposits because of their finely bedded form and the way they appear to drape over pre-existing topography. In addition, their unconformable contacts with other terrains and the widespread extent of individual layers is suggestive of eolian origins. The unlayered polar units underlie vast regions of low relief. Like the layered units, the unlayered units are also considered to have been deposited by wind. The south polar unlayered unit is thin (100–300 meters), massive, and reasonably homogeneous, but displays strong vertical, linear to gently curving structure in some areas. Layered deposits lie unconformably on both unlayered deposits and heavily cratered terrain.

The unconformable nature of the contact with the unlayered material is crucial to establishing the time relationships within the period of polar activity (see Fig. 3.27). The unlayered unit also lies unconformably on heavily cratered terrains. The size-frequency distribution of craters on the unlayered south polar deposits is like that of the oldest plains in the Tharsis region, making the unlayered polar surfaces intermediate in age among the martian plains units. The layered deposits have no indisputable superimposed impact craters, which suggests that the deposits are relatively young. (Crater counts in the polar regions, however, must be used with caution because of the possible obliteration of craters by unrecognized martian polar processes.)

The composition of the polar deposits is unknown. The remarkable preservation of craters partly exhumed from beneath polar units attests to the relatively gentle processes of burial and removal. Wind-transported particulate materials, including dust and volcanic ash, are possible constituents. Mare-like ridges near small possible volcanic mounds suggest that some consolidated rock materials also are part of the unlayered units. The incorporation of volatiles into both types of deposits has been invoked because of the polar locale, the need for a binding agent to prevent loose dust from being redistributed into deposits of variable thickness or removed entirely from the polar regions, and to explain the resistant forms of eroded slopes. Water ice is considered the prime candidate because of its stability in the martian polar environment. Carbon dioxide ice has also been considered as an appreciable component, but its incorporation in these deposits is only a remote possibility because of the ease with which it can sublime under current martian conditions.

The origin of strata within the layered deposits is a matter of speculation. A small number of resistant knobs and ledges of uneven width along the erosional escarpments suggest some variation in materials. However, lithographic variations in eolian sediments on Earth are generally limited. Moreover, the remarkably uniform morphologies of the martian deposits and their erosional surfaces argue strongly for similarly uniform material properties. The layering has been correlated with cyclic changes in the abundance of atmospheric volatiles. These changes may be induced by shifts in the eccentricity of the planet's orbit and in the obliquity of its spin axis (period $\sim$ 10^5–10^6 years) (Fig. 2.11). A possible contribution may have come from episodic production of such fine materials as volcanic ash.

Martian polar activity has been characterized by a sequence

of major climatic changes. The earliest recognized was responsible for deposition of the unlayered units; this change could have been initiated by a small change in solar luminosity. Climatic fluctuations were possibly responsible for formation of different members of these units, but local differences in source of materials or tectonic situation could also have been involved. A second change resulted in erosion of the unlayered deposits, with the formation of pits and hollows. A third change in conditions probably was needed to begin deposition of the layered deposits, which record shorter-term climatic fluctuations as well. A fourth change was required to impose the current martian environment, in which both kinds of polar deposits are under erosional attack, leading to the formation of the vast dune fields that surround the polar cap (Fig. 7.15). The dunes are composed of dark material, some of which appears to be derived from the polar deposits. Depositional and erosional activity in the polar regions thus record a pattern of complex climatic alterations over time on Mars.

Mars—The Recent Past

Some landforms reflect processes currently active on Mars. Changes in some of these features may occur on time scales varying from weeks to years, and hence have been, or reasonably should be, directly observable by orbiting spacecraft. The most ephemeral of the surface features are the light and dark markings often called "variable surface features" (Fig. 3.24). It is generally accepted that these variable albedo features result from the redistribution of fine debris by wind. One proposed mechanism calls for pervasive deposition of light material on darker units and the subsequent removal of some of the light debris to form dark streaks. Some features have changed in as brief a time as a few weeks (Fig. 7.16). Evidence of dark particulate material is seen in the images of martian dune forms (Figs. 7.15, 3.26). They apparently have changed over several months, confirming theoretical models and spacecraft observations of martian wind regimes that suggest that some dune fields are most likely active.

Figure 7.15
Part of the north polar ice caps of Mars and the dune fields surrounding it are shown in this Viking Orbiter mosaic. The dark zone is composed of hundreds of barchan and transverse sand dunes that appear to fill the low-lying region of Borealis Chasma. Detailed mapping and time-sequence images suggest that the dunes are currently active.

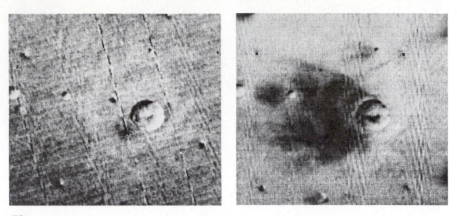

Figure 7.16
These two views of the same crater demonstrate the development of a variable feature during the Mariner 9 mission. The image at left was obtained on revolution 115; the image at right was obtained 38 days later on revolution 195. During this interval of time the small dark spot visible near the rim of the crater grew to a large fan-shaped crater streak, typical of many such features observed in the area (Daedalia region, 25°S, 125°W). The dark streaks are considered to be areas where winds have swept the surface free of loose particles, exposing a darker substrate. Area shown is about 100 kilometers wide.

Erosion of the polar deposits and deposition of debris blankets symmetric about the poles can also be classified as recent. Although direct observation of these processes cannot be made from orbiting spacecraft, their handiwork can be seen in the form of embayed, terraced escarpments; linear grooves (Fig. 7.17) presumably aligned in the direction of currently prevailing winds; subdued craters; and the dune fields discussed above.

Mass movements have played an important role in shaping the martian landscape. Some are ancient, but many show crisp scarps, fresh longitudinal and transverse ridges, and have few superposed craters. This suggests that processes post-dating the mass movements are responsible for only minor landform deterioration. Caldera walls of the Tharsis volcanoes and canyon slopes within Valles Marineris are the sites of many landslides and slumps (Figs. 3.18, 3.19). The causes of these mass movements may include undercutting of cliffs by wind; weight of overburden formed of eolian, volcanic, or impact-generated deposits near steep slopes; and possibly seismic phenomena.

Atmospheric activity involving surface or near-surface vola-

10 km

Figure 7.17
In this Viking Orbiter 2 frame of the Biblis Patera region, linear flat-topped ridges interpreted to be yardangs (streamlined, wind-eroded ridges), mesas, and lava flows dominate the scene. Mesas appear to be residual upland surfaces. Sun is from the left.

tiles should also be considered among geologic processes. Clouds, winter "hoods" that cover the polar regions, and dust storms (Fig. 7.18) are phenomena that vary on daily to annual time scales. The polar caps change dramatically during the seasons and exhibit two distinct forms of behavior. Seasonally, carbon dioxide frost caps extend equatorward to about 55° latitude and retreat rapidly and nearly uniformly. Perennial caps of water ice survive this seasonal regression and do not change appreciably.

Mars as Viewed from the Surface

On July 20, 1976, Viking Lander 1 set down on the rock surface of Chryse Planitia, becoming the first spacecraft to land suc-

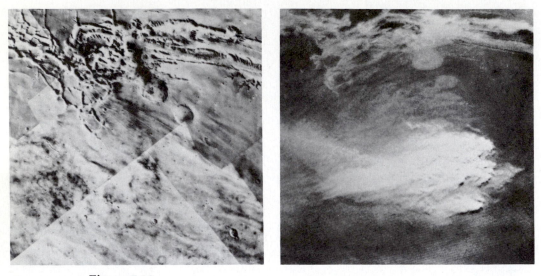

Figure 7.18
Viking Orbiter images of the Sinai Planum region of Mars south of Valles Marineris (left image), and same region during a dust storm (right image; note dust cloud).

cessfully on the Red Planet. Within seconds of landing, the camera obtained the first picture (Fig. 7.19), the first of hundreds that would provide incredibly detailed views of Mars from close up. From these, we have gained a knowledge of the surface totally unknown prior to Viking.

The area around Viking Lander 1 is strewn with rocks in the centimeter to meter size; albedo, shape, and textural characteristics indicate a wide variety of rock types. Many of the rocks are pitted; perhaps they are vesicular lavas. Bedrock surfaces appear to be exposed in some areas, and fine-grained material of probable eolian origin is visible in numerous "drifts" and "wind tails" behind rocks.

Yellowish brown is the predominant color at the site. X-ray fluorescence results, spectral reflectance data, and IR spectra from Mariner 9 and Earth-based telescopes all point toward an iron-rich chemistry for the surface and lead to the conclusion that an iron-rich clay mineral is a common component of the

Figure 7.19

This is the first photograph ever taken on the surface of the planet Mars. The center of the image is about 1.4 meters from Viking Lander 1's camera number 2. Both rocks and finely granulated material are visible. Several of the rocks exhibit apparent signs of modification by wind-transported granular material. The large rock in the center is about 10 centimeters across and shows three rough facets.

martian fine-grained particles. When the Martian surface material weathers chemically, it apparently produces a form of iron oxide called limonite, which stains everything yellowish brown.

The landing site for Viking Lander 2 was selected to be as different from that of Viking 1 as possible, yet still be safe from an engineering standpoint (Fig. 7.20). As with the search for the first landing site, the process of selection was difficult because each prospective area was viewed from orbit to be very complex and potentially hazardous. A site was ultimately selected in Utopia Planitia (Fig. 7.1), and a safe landing was accomplished. Superficially, the Utopia Planitia site looks very much like that of Viking 1, particularly because both are in plains regions and the surfaces are rocky. Closer inspection, however, shows many differences between the two sites (Figs. 7.19 vs. 7.20). Large blocks are more numerous at the second site, and the frequency of pitted rocks is greater. However, the variety of rock types at Utopia seems to be less than at Chryse Planitia, and eolian drifts are few.

A network of shallow troughs is visible at the second landing site—features totally absent at the first site (Fig. 7.21). The troughs are about 1 meter across and 10 centimeters deep; they form a crude polygonal network. Several theories of origin have been suggested, one being that the troughs are some kind of permafrost feature.

Where the landers at both sites disturbed the fine-grained material, the surface material broke into small slabs, prompting the lander scientists to apply the term "duricrust." This is a term originally applied to fine sediment in the Australian desert that had been loosely bonded by soluble minerals. Whatever the binding agent might be on Mars, the fine sediments are crusted together in the upper centimeter or so.

Many of the small "pebbles" seen at both sites actually are lumps of duricrust. Thus when attempts were made to obtain chemical analyses of "rock" samples by Viking, the rocks broke apart into fine-grained material and the attempts were unsuccessful. This result pointed to the need for a mechanism to obtain a true rock sample on a later mission to Mars.

Figure 7.20
The first picture taken by the Viking Lander 2 on the surface of Mars, within minutes after the spacecraft touched down on September 3, 1976, reveals a wide variety of rocks littering a fine-grained surface deposit. Most boulders are in the 10- to 20-centimeter size range—some vesicular and some apparently fluted by wind. As with Viking Lander 1's first picture on July 20, 1976, brightness variations at the beginning of the picture scan (left edge) probably are due to dust settling after landing. A substantial amount of fine-grained material kicked up by the descent engines has accumulated in the concave interior of the footpad. Viking 2 landed at a region called Utopia in the northern latitudes, about 7500 kilometers northeast of Viking 1's landing site on the Chryse plain 45 days earlier.

Figure 7.21

Viking Lander 2 photographed this rocky panorama shortly after touchdown on September 3, 1976, on the northern plains of Mars in the Utopia region. The picture shows all but about one-tenth of the angular view from the lander. The surface is strewn with rocks all the way to the horizon. The largest are several meters in diameter. The pitted rocks resemble fragments of volcanic lava, which are frequently quite porous. Some of the rocks appear to have grooves that may have been cut by the impact of windborne sand and dust grains. There is no indication of sand dunes in the scene, although deposits of fine-grained material occur beneath and between boulders.

Changes in the landscape were observed over the more than two (terrestrial) years of observations. Wind-generated changes, minor slump, and a peculiar white "frost" were among the transitory phenomena observed.

With each succeeding mission to Mars, the planet is found to be more complex than previously anticipated. This was certainly true with the views of the surface in regions thought to be rather bland and smooth prior to Viking. The origin of the large rocks, the character of the pitting, and the complex surface chemistry have all piqued the minds of planetary scientists.

Martian Microbes?

What is "life"?—and once life is defined, how do you find life on another planet? Those were the questions facing the biologists of the Viking mission. For the years leading up to the landings on Mars, experiments, procedures, and hypotheses were tested in Earth-bound laboratories in anticipation of the Viking results. After the landings, the most clearly established result of the Viking experiments dealing with the search for life is that the soil of Mars in the areas tested differs profoundly from that of Earth. Mars is truly alien by virtue of the complete absence of organic materials. Even the most inhospitable soils on Earth, such as those of the dry valleys of Antarctica, contain more organic material, directly or indirectly connected with present and past life, than do those in the areas tested on Mars.

The outcome of the Mars search for life by Viking may be regarded as having ruled out terrestrial chemical and biochemical conditions and therefore very probably Earth-like life. But it could not rule out the existence of a biology uniquely adapted to the alien chemistry that characterizes the martian soil.

Mars remains a most fascinating planet, even though it is probably lifeless at present. Where did its water go? What are the origin and nature of the climatic fluctuations that are so dramatically revealed in the sedimentary deposits of its polar areas? Is Mars still volcanic or otherwise seismically active? These and other questions remain to be tackled by surface exploration in one form or another, supplemented with orbital measurements. Indeed, a global geochemical mapping by remote sensing is an essential scientific milestone. Later, truly autonomous rovers operating over thousands of kilometers of landscape may become invaluable exploratory tools. Ultimately sample return itself from Mars must be a paramount scientific goal.

So the question is, "Where do we go from here?" Should the "Search for Life on Mars" thrust of the United States planetary program continue to be of central focus? That is a question which must and will be addressed. However, the scientific focus after Viking will inevitably be more chemical than biological. Furthermore, the special public interest created by the search for life on Mars may, to a large extent, have been dispelled by the absence of any large organisms or other biological evidence visible in the pictures. Any further search for life on Mars will be part of a more balanced program of exploration of that planet.

SUGGESTED READING

Batson, R. M., P. M. Bridges, and J. L. Inge. *Atlas of Mars,* NASA SP-438, 1979.

Carr, M. H., and N. Evans. *Images of Mars—The Viking Extended Mission.* NASA SP-444, 1980.

Leighton, R. B., N. H. Horowitz, B. C. Murray, R. P. Sharp, A. H. Heriman, A. T. Young, B. A. Smith, M. E. Davies, and C. B. Leovy. "Mariner 6 television pictures: First Report." *Science* **165**:684–690, 1969.

Leighton, R. B., N. H. Horowitz, B. C. Murray, R. P. Sharp, A. H. Heriman, A. T. Young, B. A. Smith, M. E. Davies, and C. B. Leovy. "Mariner 7 television pictures: First Report." *Science* **165**:787–795, 1969.

Leighton, R. B., N. H. Horowitz, B. C. Murray, R. P. Sharp, A. H. Heriman, A. T. Young, B. A. Smith, M. E. Davies, and C. B. Leovy. Mariner 6 and 7 television pictures: Preliminary Analysis. *Science* **166**:46–67, 1969.

Mariner 9 Preliminary Science Report. *Science* **175**:293–322, 1972.

Mariner 9 Interim Science Report. *Icarus* **17**:289–517, 1972.

Mariner 9 Final Science Report. *J. Geophys. Res.* **78**:4007–4440, 1973.

Mutch, T. A., R. E. Arvidson, J. W. Head III, K. L. Jones, and R. S. Saunders, *The Geology of Mars*. Princeton: Princeton Univ. Press, 1976.

Proceedings of the First International Colloquium on Mars. *J. Geophys. Res.* **79**:3375–3410; 3888–3971, 1974. *Icarus* **22**:239–396, 1974.

Proceedings of the Second International Colloquium on Mars. *J. Geophys. Res.* **84**:7909–8544, 1979.

Viking Preliminary Science Reports. *Science* **193**:759–815, 1976. *Science* **194**:57–105; 1274–1353, 1976.

Scientific Findings from Mariner 6 and 7 Pictures of Mars: Final Report. *J. Geophys. Res.* **76**(2):293–472, 1971.

Scientific Results of the Viking Project: Final Reports. *J. Geophys. Res.* **82**(28):3959–4680, 1977.

Viking 1 Early Results. NASA SP-408. Washington, D.C.: U.S. Government Printing Office, 1976.

Viking Lander Imaging Team, *The Martian Landscape*, NASA SP-425, 1978.

Note: The U.S. Geological Survey is currently publishing a series of geological maps of Mars at a scale of 1:5 million as part of its Miscellaneous Investigations Map Series.

8

COMPARATIVE PLANETOLOGY

COMPARATIVE
PLANETOLOGY

Darwin's nineteenth-century theory of biological evolution set in motion an intellectual revolution that continues to shape modern thinking. We humans are a *part of nature, not apart from it.* The burst of new data and ideas about the inner Solar System that has accompanied and been stimulated by the first two decades of the Space Age is really a continuation of the process by which we have come to understand our affinities with our physical environment. Our planetary neighbors are not alien; they are Earth's planetary kinfolk. Earth arose from the same "cosmic stuff"—at about the same time and under the influences of similar processes. Only modest differences in orbital distance from the Sun seem to have determined whether a watery, habitable planet formed rather than one with the uninhabitable environmental extremes of Mars or Venus.

The intellectual revolution stimulated by the Space Age has enlarged our consciousness of the Earth and its planetary neighbors. The Moon has become almost a familiar place. Mars no longer harbors the fanciful beings of Ray Bradbury, Edgar Rice Burroughs, or Robert A. Heinlein, but, instead, has been revealed as a stranger place—albeit lifeless—than any novelist could have imagined. Mercury is at last real—part of scientific reality, if not yet of popular perspective. Venus is superficially bizarre, but we are growing accustomed to such tendencies in our close planetary relatives, especially as we realize how fundamentally alike are the chemical histories of these planets.

Scientific consciousness has been expanded to encompass the realities of the inner planets, not just exciting speculations. Popular reality is growing to recognize the global interconnectedness of all environments on Earth and, especially, the uniqueness of this one habitable planet in our Solar System.

Much of the early scientific motivation for photographing the surfaces of the inner planets and for analyzing samples of the Moon and meteorites was to pierce the veils surrounding planetary birth—ours and that of our neighbors. In this searching re-examination of planetary genesis, many traditional "truths" have been found to be inconsistent with new facts, and a broad range of new ideas has been generated to try to describe what we now recognize to have been complex and diversified events that accompanied the first few hundred million years of the Sun's existence, following the condensation of the solar nebula around 4.6 billion years ago. Yet the obscuring veils really have not been removed, just opened here and there enough to permit a glimpse of astonishing circumstances only dimly understood.

Instead of finding clear records of planetary birth on the surfaces of our neighboring planets, the space effort revealed a wholly unanticipated era of early planetary evolution, ending about 4 billion years ago. During this early stage of evolution, violent internal and external processes fashioned all the earthlike planets; their subsequent evolution has been comparatively quiet. Moon and Mercury were senescent long before the first multicellular organisms evolved on Earth, less than 1 billion years ago. Most of the major surface features on Mars, except in the polar regions, probably formed during those "middle years" of planetary history, between 1 billion and 4 billion years ago.

The next few sections summarize and compare this newly revealed middle history of the planets. The final section presents a look ahead into what the future may hold for further exploration.

THE LATE HEAVY BOMBARDMENT PERIOD

The Surprise Testimony of the Lunar Rocks

Perhaps the greatest scientific surprise of the Apollo program was that the scarred topography of the lunar highlands—previously considered to be left over from the accretionary process—is significantly younger than had been anticipated.

Indeed, it was formed more than half a billion years after the main planet accumulated. The Moon's surface had cooled and had recorded unknown earlier landscapes long before the present topography was formed. Furthermore, the age estimates gained by intricate study of the isotope ratios in the mineral grains of returned samples, as well as from photogeologic studies, suggest a relatively rapid cessation of the final episode of bombardment recorded on the highland surface about 4 billion years ago. It has not been possible to determine confidently whether that bombardment episode was an abrupt tailing off of accretion—that is, impact by objects in Earthlike orbits—or an entirely distinct episode, corresponding to the influx of objects from farther out in the Solar System, perhaps even bringing in volatiles. Either way, a period of late heavy bombardment occurred not just on the Moon, but on Mercury and Mars as well (Fig. 8.1). And, of course, it must have played an important role on Earth.

Thus the oldest record that planetary geologists have confidently extracted so far as a result of correlating the geological records from the surfaces of the inner planets has been this important epoch of massive bombardment. It affected the entire inner Solar System, and we refer to it as the Late Heavy Bombardment Period.

Survival of Scars from Late Heavy Bombardment on Mars (and Venus?)

On Mars, the Late Heavy Bombardment Period probably coincided with a period of significant atmospheric density, which came to an end shortly thereafter, the loss of the constituent gases being due either to gravitational escape or to absorption and retention by a thick global layer of finely divided material. The surface abundance of volatiles was high very early—but only very early—in the history of Mars. Most of the enormous, channel-like features that have attracted so much attention apparently formed shortly after the end of this heavy bombardment, during a fluvial (but not oceanic) period, probably between 3 and 4 billion years ago. The next phase was characterized primarily by development of widespread basaltic plains that filled previously formed basins as well as other portions of the surface.

Whether Venus likewise retains a topographic record from the period of late heavy bombardment is a question of extreme interest in understanding the history of that planet. There are, indeed, indications of large, circular topographic features in the images obtained by powerful Earth-based radars. These could easily be residual scars from that impact epoch. However, volcanic interpretations are also possible for these features. Earth-based radar is limited to very low surface resolution; space-borne instruments will be required for significant progress.

Earth—The Missing Chapter

What about the Earth itself during this early epoch of Solar System history? Interestingly, no Earth rocks have been measured with confidence that are older than about 3.8 billion years. Nor has any evidence of impact been recognized in these oldest rocks. Can the catastrophic effects of late heavy bombardment have nearly destroyed and reworked all previous geological evidence on the Earth? What was the effect on the atmosphere and hydrosphere that may have existed at that time? Is it even possible that the bombardment process delivered to the Earth the volatiles that became ocean and air and the constituents of the life-forming molecules that arose early in Earth's aqueous history?

The meager available geologic record suggests that the first self-replicating molecule—the beginning of life—may have originated about 3.5 billion years ago, shortly after the bombardment epoch. It is as if the Earth underwent a major transition at that time—perhaps its most important transition from the point of view of the surface. At any rate, the basic elements that are considered important to the evolution of the hydrosphere, the atmosphere, and perhaps even the earliest life processes on Earth were certainly present.

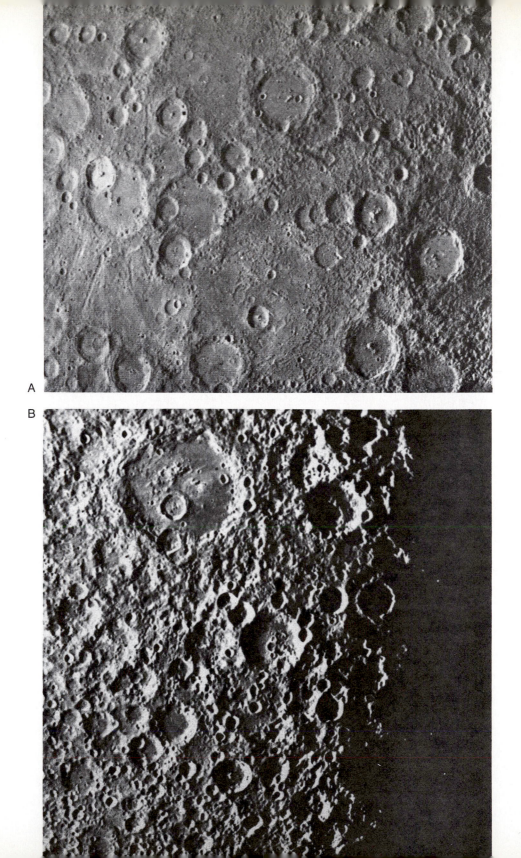

A

B

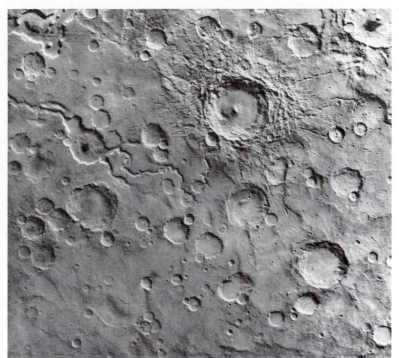

C

Figure 8.1

Comparison of cratered terrains on Mercury, Moon, and Mars. **(A)** Mariner 10 photograph of the bright-rayed crater Kuiper and surrounding cratered terrain on Mercury centered at 13°S, 25°W. The solar elevation angle varies from about 5° to 27°. **(B)** Mariner 10 photograph of a region in the northern uplands of the Moon centered at 60°N, 130°E. The solar elevation angle varies from 0° (i.e., the terminator) to about 25°. **(C)** Viking Orbital photograph of cratered terrain in the Deuteronilus region on Mars, centered at 30°N, 340°W. The solar elevation angle varies from about 11° to 16°. The left-to-right dimension of the Mariner 10 frames is about 900 kilometers. All three images have been projected at the same surface scale. Both Mars and Mercury exhibit extensive intercrater plains, deficient in 10- to 50-kilometer-craters in comparison with the Moon. The Mercurian craters, however, resemble in distribution and morphology those of the Moon rather than the highly degraded population present on Mars. [After Bruce C. Murray, Robert G. Strom, Newell J. Trask, and Donald E. Gault, *Journal of Geophysical Research*, vol. 80, p. 2510, 1975. Copyrighted by the American Geophysical Union.]

MARIA ON THE MOON, MARS, AND PROBABLY MERCURY

A major lunar geological event was the filling of the basins by the dark maria, and that fact was recognized before the space age. Space age data indicate that similar epochs also occurred on Mercury and Mars, despite their very different internal geophysical conditions and locations within the Solar System. Indeed, the features on some martian and lunar plains are so similar that high-resolution orbital photographs of them are practically indistinguishable (Fig. 8.2). Although the mercurian plains exhibit small but significant differences from those of the Moon and Mars, the most probable explanation for them also is that they were formed by massive floods of lava that filled the basins remaining from the Late Heavy Bombardment Period.

Is there a connection between the formation of the basins and that of the basalt plains that filled them? On the Moon, the formation of the Imbrium Basin and the emplacement of the surface lavas there were separated by several million years. There does not seem to be any direct way for basin-scale impact to serve as the cause of massive basaltic flooding hundreds of millions of years later. Yet we cannot ignore the similarity of relationships between basins and subsequent lava fillings on Mars, the Moon, and perhaps Mercury. Was the state of the interior on Mars independently (and coincidentally) so similar to that of the Moon that delayed basaltic response would also occur? Or was basaltic volcanism occurring so frequently on all three bodies that there really is no comparable time gap at all? Future age measurements will help determine how "coincidental" the events really were on the inner planets.

The Maria of Mars

Impact crater abundances suggest (but do not prove) that the most ancient basaltic plains on Mars are probably about 3 billion years old. They are younger than most major martian channel features and, in fact, sometimes bury them. Thus the channels must have been formed primarily between the end of the heavy bombardment and the emplacement of the basalts, during a narrow time span associated with the demise of an early dense atmosphere and the onset of atmospheric conditions perhaps not much different from those of the present. However, some channel formation may well have continued into the epoch of plains volcanism. Yet many impact craters on these old martian basalt plains appear as fresh in many ways as lunar ones and do not show the kinds of degradation features that would be inevitable if there had been subsequent episodes of atmospheric density over martian geologic time. Evidently, the "aqueous" or heavy-atmosphere phase of Mars appears not to have survived much beyond the end of heavy bombardment, although some channel formation may have continued beyond the demise of that early atmosphere.

It is difficult even to guess whether there was time enough then for even the simplest form of life to evolve. Certainly there was never as much time for biological synthesis on Mars as there was on Earth, nor did Mars ever have Earth's diversity of favorable environments. Present martian conditions are quite hostile, not only to life but to organic material in general, and these conditions may have prevailed for a very long time. The most probable inference that can be made about the possibility of life developing on primitive Mars is that the environment over much of Mars' history seems to have been less than favorable. Yet we cannot definitely conclude on this basis alone that there never has been life on Mars.

Did the Earth Evolve Through a Moonlike Phase?

What about Earth? Were massive plains basalts emplaced on our planet as well? Large masses of plains basalts do, indeed, make up parts of the Earth's continents. However, they are much, much younger than those of the Moon or Mars and certainly do not correspond to the filling in of ancient impact basins. Indeed, the process on the Earth that creates the greatest amount of basalt is that which forms the sea floor. The upwell-

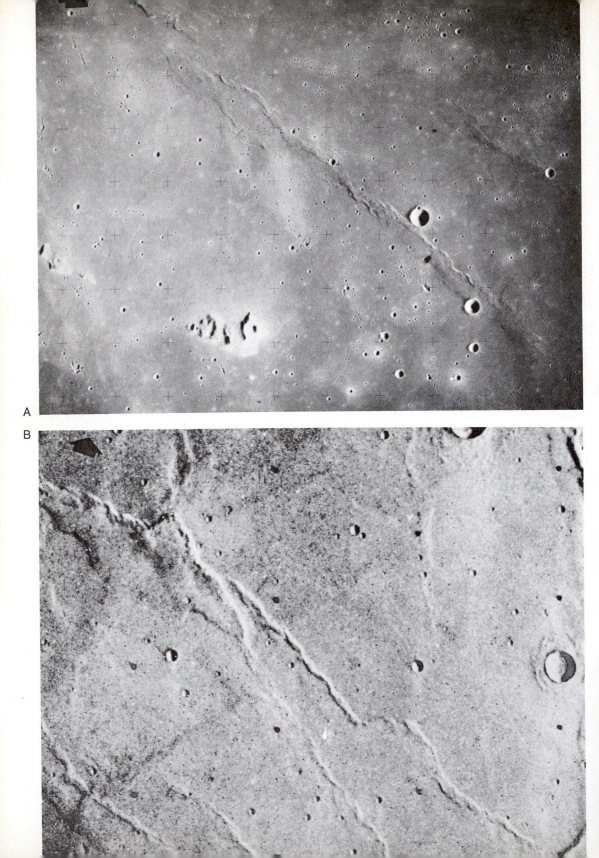

A

B

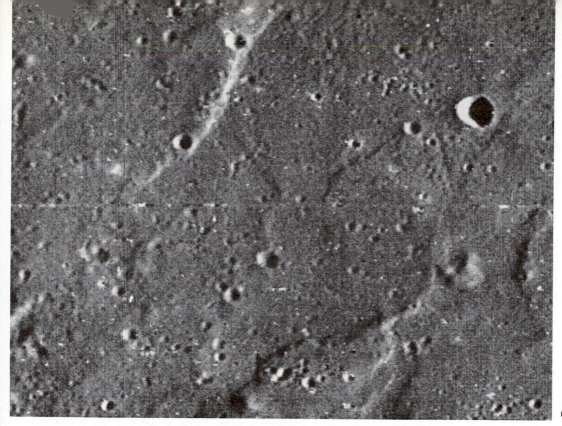

C

Figure 8.2

Mare ridges on Moon, Mars, and Mercury are compared in these three frames. **(A)** Apollo 15 metric view of part of Mare Imbrium, showing typical mare ridges. Mons La Hire is the mountainous block in the lower left quadrant. **(B)** Viking Orbiter image of part of Hesperia Planum on Mars, showing mare-like ridges. **(C)** Mariner 10 view of southern Suisei Planitia (east of the Caloris Basin) showing mercurian plains ridges. All photographs are at the same scale (about 100 kilometers by 120 kilometers), with comparable illumination and viewing angles; north is toward the top; resolution, however, ranges from excellent for the Moon (less than 50 meters) to poor for Mercury (about 500 meters).

ing of fresh basaltic material takes place at the mid-oceanic ridges. This material spreads out to form new basaltic crust as other parts of the Earth's crust recede to be destroyed at plate boundaries. Thus subsurface activities causing plate motions on the surface are also constantly delivering large amounts of new basalt to the surface, but not in the same way that it has reached the surfaces of the Moon, Mars, and probably Mercury.

Early in its history, did the Earth ever bear a resemblance to the Moon or the Mars of today? At one time, the answer might have been "yes," for the view was once held that the Earth accreted as a cold body that gradually heated up. The plate-tectonics mechanism that now dominates the surface was thought to have arisen midway or later in its history than the tensional stresses that produced tectonic features preserved on the other planets (Fig. 8.3). However, the present view of many specialists is that Earth's lithosphere is now and always has been too mobile to preserve large impact basins and subsequent mare-like filling over hundreds of millions of years. Rather, the Earth has been gradually cooling over its history, and mantle convection from within has been modifying its surface throughout that time.

Ancient Maria on Venus?

What about Venus? Is it like the Earth, or are there still evidences of a Moonlike history at some point? Earth-based radars provide tantalizing indications of large, linear depressions that could be interpreted as rift valleys or other features similar to those produced by plate-tectonic phenomena on the Earth (or as features like Valles Marineris on Mars). However, the resolution of radar is not good enough to allow confident interpretations. A major scientific objective in the future study of Venus is to acquire directly comparable "images" of topography so that plate-tectonic phenomena as well as other volcanic and structural features, if present, can be unambiguously identified.

We Earthlings are fortunate to have available for our study two extremes of atmospheres among the neighboring planets to help us understand our own atmosphere. Venus, whose atmosphere is one hundred times denser than ours, exists in a permanent "hothouse." Mars, whose atmosphere is one-hundredth as thick as Earth's, is in an ice age. Similarly, nature has facilitated our study of Earth's magnetic field by supplying Mercury with a miniscule but morphologically quite similar field. Perhaps, in addition, nature has been good enough to provide another planet, Venus, with a mobile, thin lithosphere to help us understand the extraordinary phenomenon of plate tectonics, which controls the Earth's major geologic and topographic features, as well as much of the localization of its natural resources.

On Earth it is difficult to separate the profound effects of liquid water on the chemical distribution of surface materials from the effects of internal activity. If Venus is found to lack surface water but to be internally active, it will help greatly to sort out the causes of regional chemical differences on Earth.

THE MIDDLE YEARS

The Moon and Mercury have remained virtually quiescent, insofar as we can tell, since the emplacement of the basalts 3–4 billion years ago. A light rain of asteroidal and cometary debris has peppered their surfaces to a remarkably similar degree. No atmosphere has modified their surfaces, and thus those surfaces are fossils from the birth and early development of the planetary systems.

Magnetism, Magmatism, and the Interior of Mercury

The existence of Mercury's Earthlike magnetic field remains an enigma. The simplest explanaton of the mercurian field is that it is, indeed, produced by a fluid iron core currently undergoing convection within the planet. For this to happen, the interior heat must either be prevented from escaping to the surface

A

B

Figure 8.3

Graben systems on the planets, indicative of local regions of crustal extension. **(A)** Earth: Low altitude aerial oblique view of the Canyonlands graben system, southeastern Utah. Rocks are extensively jointed parallel to graben strike direction. Width of foreground is a few kilometers. **(B)** Mars: Grabens near the caldera of Alba Patera (40°N, 115°W). Oblique

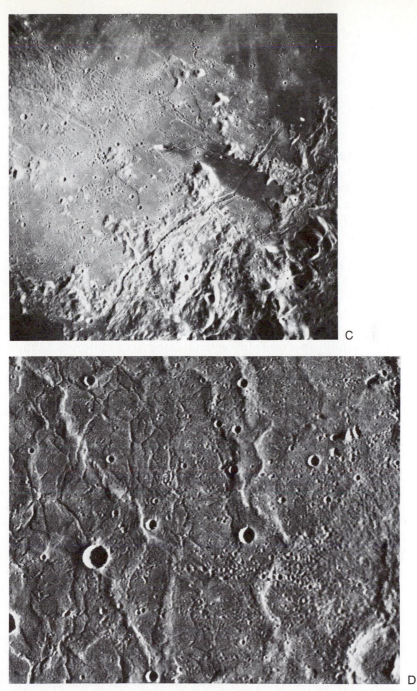

view; width of foreground is approximately 100 kilometers. **(C)** Moon: Graben (20°N, 0°E) parallel to Appenine Mountain scarp, concentric to the Imbrium Basin. Area shown is approximately 100 kilometers by 100 kilometers. **(D)** Mercury: Grabens in floor materials of the Caloris Basin (33°N, 180°W). Width of area shown is approximately 200 kilometers. [A, courtesy of G. McGill.]

or be replenished somehow at a sufficienty high rate to maintain the fluidity of the iron core. A future orbiting satellite is required to unravel this paradox of how a small, slowly rotating planet can apparently so closely mimic the complex internal dynamics of Earth.

Continued Basaltic Volcanism on Mars

Unlike the Moon and Mercury, Mars shows a further development of volcanism in the form of gigantic shield volcanoes and associated large-scale fracture systems. These huge piles of basaltic lava probably accumulated over hundreds of millions of years through continued eruption of magma to the same surface area. This does not happen on Earth. Because Earth's plates are constantly moving, they remain over the "hot spots" that produce basalt only for comparatively short periods. The Hawaiian Islands, for example, consist of a whole chain of volcanoes that have formed on the ocean bottom as a result of the passage of the Pacific Plate over an inferred mantle "hot spot." Mars apparently has no crustal blocks that move relative to the lower layers. This is a principal difference between basaltic volcanism on Mars and Earth. The lack of plate tectonics on Mars is further indicated by the absence of folded mountain belts and other characteristic features of the Earth.

The Moon, Mercury, and Mars all probably have much thicker lithospheres than does the Earth. The lithosphere (not to be confused with the chemically defined *crust*) is the cool and relatively rigid outer layer of the planet which acts as a uniform elastic solid layer. The Earth's lithosphere apparently rests on a relatively "slippery" substratum and is thin enough so that it has broken into plates that are moved around by planetary-scale convection in the underlying mantle material. Convection conceivably takes place on Mars as well. Indeed, the "lumpy" global surface shape and gravity field of that planet may be related to deep convection, but such mantle movements have not been strong enough to fracture and crumple the lithosphere, as has been the case on Earth. Hence the upwelling magma has simply continued to accumulate in a few places, and the outer lithosphere has retained its integrity throughout many billions of years. The Moon and presumably Mercury also have thick lithospheres—thicker than those of Earth or Mars. There is no evidence that convection has even deformed, much less fractured, the surfaces of either planet for billions of years—if, indeed, convection is still extant. This pattern is consistent with their smaller size (and higher surface-area-to-mass ratio) and therefore greater effectiveness over geologic time in radiating internal heat away to space.

Thus Mars now seems to be at the waning stage of a long period of gradually declining volcanic activity, although that conclusion could be modified by future data.

Mars, Earth, and the Challenge of Climatic Changes

The martian atmosphere, after having made the transition from dense to thin shortly after the end of the Late Heavy Bombardment, most likely has remained relatively thin ever since. However, major climatic fluctuations of that "thin" atmosphere are clearly indicated by the layered deposits in the polar regions. Such fluctuations likely took place over hundreds of thousands to millions of years; such periodic variations in the amount of sunlight reaching Mars are to be expected as a result of the perturbations in spin-axis orientation and extremes of heliocentric distance which that planet undergoes. In addition, cyclical or secular changes in the luminosity of the Sun must somehow be recorded there. Mars' polar regions probably contain unique, detailed records of small fluctuations in eolian sedimentation that in turn reflect small variations in solar radiation that must also have affected the Earth. As Mars' polar regions are eventually explored in detail, both from orbit and in situ, they may well provide vital information relevant to the origin of drought-causing climatic fluctuations on the Earth, as well as the origin of the terrestrial glacial epochs themselves.

Indeed, humans exist now in what probably is a brief epoch within the periodic glacial stages that have dominated the past

3 million years. Similar extensive glaciation apparently occurred only twice before that time in the last billion years. There is no obvious explanation why the dramatic glacial phenomena occur only every few hundred million years. One possibility is that chance collisions of crustal plates have profound effect upon the oceans and atmosphere by causing massive evaporation and changes in ocean currents. Apparently the Mediterranean Sea was squeezed off from the Atlantic Ocean about 15 million years ago and literally evaporated. Some scientists believe this might have had major climatic effects and caused the subsequent onset of glaciation. Alternatively, excess volcanic activity could have intensified that stratospheric dust cloud, thereby reducing the amount of sunlight reaching the Earth and thus leading to worldwide cooling.

On the other hand, glacial ages may be related primarily to the modification of the heat balance by unusual solar activity rather than chance events on the Earth's surface. External changes in solar radiation would be recorded on Mars as well as on the Earth. Examining data from Mars can help determine the effect of variations in sunlight on Earth's glaciation. Information about climatic change on Mars and the part of its history that corresponds to Earth's glacial epoch and the subsequent warming trend should be a primary goal of future martian exploration.

The Promise of Venus

Russian landers on Venus—Venera 9 and 10—accomplished the very difficult feat of taking facsimile pictures of the terrain immediately surrounding those landers and radioing them to the Earth. The pictures show evidences of seemingly fresh erosion and renewal on the surface of that planet. Thus Venus does seem to be an active planet, even at present. Therefore, it probably has a long geologic record of activity—perhaps as extensive as the Earth's record. Investigation of the geologic history of Venus remains one of the great exploratory tasks for the future.

WHERE NEXT?

The completion of the Viking mission to Mars and of the 1978 Pioneer Venus probes and orbiter of Venus mark the conclusion of a great burst of American space exploration of the inner planets. New missions are being contemplated, but the level of activity will be at a much more modest pace for some years or decades to come. What should be the objectives for renewed exploration? Can we expect to gain as important new insights in the future as have characterized the remarkable pioneering space ventures of the 1960s and 1970s?

The Utility of the Moon

For the Moon, the guiding motivation in the future may well be utilization rather than exploration. When new lands on the Earth were first being explored, much of the initial interest was in understanding what was there. This was followed by an inevitable interest in using the resources of the new territory. At first, the prospect of using resources from the Moon may seem too distant to warrant twentieth-century attention. However, it may be instructive to recall that the first exploration of the Antarctic continent took place less than eight decades ago, and was characterized by a nationalistic race to the pole. After two world wars and the development of much more effective transporation technology, the 1957 International Geophysical Year took place, during which large-scale scientific exploration of Antarctica was undertaken, primarily for scientific reasons. Now, as global energy and resource shortages become more serious (and offshore oil drilling and other operational techniques have been perfected), the exploitation and utilization of resources, at least at the edge of the Antarctic continent, are becoming probable. Thus, even for the remotest part of the

Earth's surface, the time scale from esoteric exploration to first utilization may well prove to be about a century.

The Moon has a potential significance far greater than just as an object of scientific interest in its own right. From the earliest days of space exploration, the Moon has been envisioned as a potential base for manned space operations. Indeed, the establishment of large-scale automated radio-astronomy systems on the Moon, especially on the farside, seems almost a certainty. The reason is that the Moon's farside is uniquely free of interference from terrestrial radio signals that even now limit the investigation of natural radio emissions from the cosmos. Indeed, the rapid development of satellite-to-ground transmissions for position determination, communications, and other purposes is rapidly making large, virtually unexplored frequency regions of the cosmic electromagnetic radiation unobservable from Earth.

Furthermore, the Earth's ionosphere is a natural emitter of radio emissions in the kilocycle and low-megacycle region of the radio-frequency spectrum—an inevitable side effect of its use as a natural reflector for conventional AM radio signals. It simply is not possible to perform radio astronomy in these low-radio-frequency regions from the Earth's surface or even from stations orbiting the Earth. In addition, Earth-orbiting radio telescopes are vulnerable to unintentional radio-frequency interference, since the Earth's ionosphere is transparent to a large part of the electromagnetic spectrum. As the terrestrial communications revolution continues, the natural cosmic radio background becomes increasingly polluted. But the far side of the Moon always faces away from the Earth and is shielded from all terrestrial radio emissions—hence its unique value as a site for celestial radio observations.

Recently the interest in the search for evidence of extraterrestrial intelligence has begun to focus seriously on listening for microwave signals of artificial origin created by aliens living on planets orbiting other stars. Someday, when the human imagination is matched by adequate resources to carry out a full-scale radio investigation in search of alien intelligence, the far side of the Moon may well prove to be an especially precious and important environment.

Similarly, as space transportation becomes ever cheaper and

more effective, we must anticipate the utilization of lunar resources. They may be used on the Moon itself, or, in extreme cases, transported to Earth orbit for use there, or even brought to the Earth's surface. The initial efforts could be made within the lifetime of our children and probably will become part of the activities of our children's children. Hence further scientific study of the Moon, with an eye toward identification of its resources and their utilization, provides a logical target on which to focus near-term studies. For example, the illumination of sunlight in the polar areas is nearly constant in intensity and variable merely in azimuth. Both automated and manned endeavors might be undertaken in these regions, especially if sunlight is used as a source of electrical power and heat. Similarly, frozen volatiles, if they exist on the Moon at all, may be found in permanently shaded areas at the poles. Hence we can already suggest small regions of the Moon to explore in much greater detail (regions that were not examined in the course of the Apollo program because of their inaccessibility then) as leading candidates for future utilization.

On an even longer time scale, the exploitation of the Moon will probably become a human endeavor. Already there is serious discussion by a few visionaries concerned with the creation of special self-sustaining "colonies" at gravitationally stable points of the Earth/Moon system. But in order to do this, it is presumed that the Moon will have to serve as both a staging base and a source of raw materials. The eventual possibility of space manufacturing, and especially the possibility of acquiring additional energy resources by means of large stations in orbit about the Earth, is no longer solely the province of science fiction writers. Serious feasibility studies are under way. As a consequence, it is plausible that at some point in the next century the Moon will become the focus of more than just scientific research for its own sake—that it will become a site of actual human activity serving practical human needs on Earth, and that its natural resources will be exploited.

The Moon will probably figure large in our future and should be an object of further scientific exploration in the nearer term. Further automated—and eventually manned—scientific activity can be regarded as the likely extension of the technical capability already developed on Earth.

Return to Mercury

Observations by orbiting satellites of Mercury's extraordinary Earthlike but miniature magnetic field could provide very important information about the mysterious magnetic field of the Earth itself. And, of course, half of its surface remains totally unexplored.

Mercury, however, is intrinsically a fickle place to explore, both because of the hostile environment arising from the intense solar radiation and, most importantly, because of the high speed with which a terrestrial spacecraft normally would approach the planet. It was only by substituting cleverness for chemical rocket propulsion that the Mariner 10 mission was possible at all. A close flyby of Venus was used to perturb the trajectory toward Mercury. In addition, the resonant nature of the resulting orbit with that of Mercury was exploited to provide two additional passes by the planet. Similar cleverness probably will be required in the future to slow down a spacecraft and permit it to be captured by Mercury's gravity field so it can orbit that planet.

Fortunately, there is yet another good new idea that may permit brain power to be substituted for the enormous rocket power required, both to go to Mercury and then to orbit that planet. This is the concept of low-thrust continuous propulsion, which could even be used for a return trip if desired. The technology most likely to make low-thrust propulsion to Mercury a reality is "ion drive" (Fig. 8.4). Here sunlight is absorbed by large solar-electric cells in huge arrays. The resulting electrical power is used to run large ion engines. Thrust is obtained in an ion engine by accelerating individual ions to very high velocities and discharging them through nozzles (Fig. 8.5). (Solar sailing is another distant possibility, especially after construction of large space systems in Earth orbit become a reality.)

Ion drive is not a new concept. Research and development on small-scale systems has been underway for nearly two decades. However, it has only been with the advent of the United States' space transportation system, "the shuttle," which combines large-volume capability with various advances of space technology, that a high-capacity interplanetary system could even

Figure 8.4
Sunlight would be converted directly into electricity in an ion engine by large silicon-cell arrays deployed like giant wings. The resultant electrical power would be used to accelerate to high velocity ions of mercury or another substance out of a battery of rocket nozzles (lower center). The resultant low acceleration (approximately one-thousandth Earth's gravitational acceleration), when applied continuously, could carry the payload and ion-drive propulsion system to rendezvous with comets and asteroids as well as the planets—and even allow automated return of collected samples.

Figure 8.5
Four prototype 20-centimeter ion engines are shown mounted in a test frame. The frame is attached to the closure end of a vacuum-chamber test facility. A volleyball-size propellant tank holds enough liquid mercury propellant to run all four engines for thousands of hours. Ground tests in vacuum chambers allow investigation of ion-beam optics, engine-to-engine interactions, operational characteristics, and lifetimes.

be contemplated. Thus future space travel to Mercury no longer seems especially difficult. In fact, if ion drive is developed as a planetary propulsion system by the United States, missions to Mercury would be among the easiest to accomplish because of the great increase in performance obtained by going near the Sun. Indeed, even *sample return* from Mercury soon may be regarded as technically feasible.

What about the distant future? In the case of the Moon, Mars, and Venus, the major step following remote-sensing missions has been the landing of an automated vehicle. This could be done on Mercury later in this century. As technology was developed for the earlier missions, capabilities increased without comparable additional cost. For example, the Viking mission to Mars was by far the most sophisticated automated space mission, with the demanding goal of searching for alien microbial life there. Yet its cost in absolute dollars was comparable to the Surveyor missions to the Moon 12 years earlier, when only simple mechanical and elementary chemical tests were carried out. Hence, one or two decades from now, simple measurements on the surface of Mercury should be relatively inexpensive by comparison. And even more ambitious missions are possible, such as automated return of a sample to the Earth for analysis, thereby opening the possibility of absolute age dates and interplanetary age correlation. Mercury is a most valuable scientific frontier for the future.

Will Mercury ever become interesting from the viewpoint of exploitation? Certainly not until the exploitation of the Moon has become a reality. On a very extended time scale—and considering the growing importance of energy to an overstrained and overworked Earth—the fact that each square centimeter of Mercury's surface receives five to ten times as much solar energy as does its counterpart on the Moon conceivably could become significant. Like the Moon, Mercury also has permanently shaded polar areas that create enormous temperature differences capable of producing useful power, and the shaded areas similarly may contain volatiles.

Furthermore, Mercury provides at least a weak magnetic field that should partially shield any human visitors from harmful high-energy solar and cosmic radiation. Thus even Mercury may hold interest as a site for human activity in the twenty-first century.

Geographic Exploration and the Continued Lure of Mars

One of the many surprises to come out of the Viking mission was the recognition of the very significant gap between the smallest surface features that could be resolved from orbit and those found to dominate the landscapes surrounding both Viking landers. From orbit, the planet appears to be remarkably heterogeneous, with a wide variety of landforms. Yet the terrains surrounding the two Viking landers are in many ways similar despite the deliberate effort to place them in different microenvironments. Of course, both the target areas were chosen in the expectation of minimizing the landing hazards created by large boulders or sharply sloping surfaces. (Even so, one leg of Viking 2 rested on a boulder large enough to tilt the spacecraft eight degrees.)

It is very clear from that experience that future landing missions will require much greater accuracy in targeting in order to place the landers in or adjacent to especially interesting areas, and that robots delivered to the surface should be mobile. Furthermore, large-scale mobility, permitting the surface to be explored over hundreds to thousands of kilometers, will require "smart" robots. Because tens of minutes are required for round-trip communication with Earth, it will not be practical for an Earth-based operator to directly operate a rover on Mars.

On the other hand, the next step may have to be more modest than was Viking. Exploring Mars is expensive, and national priorities may require an extended period of time before a major new Mars initiative can be undertaken. There are still high-yield low-cost opportunities to learn more about specific locations on Mars. For example, carefully targeted "penetrators" (high-velocity surface-penetrating probes) or hard landers are possibilities. Even a Mars automated glider or airplane is being studied! All of these would probably be best delivered and supported by an orbiting vehicle that could provide further observations from orbit to refine and extend the Viking and Mariner 9 data.

On a longer time scale, it seems likely that a truly autonomous, specially designed vehicle will be developed—one capable of navigating through rocky terrain on its own with relatively little "advice" from Earth-based engineers and traversing thousands of kilometers (Fig. 8.6). Such a system might be de-

Figure 8.6
Futuristic free-wheeling labs on Mars are illustrated in this artist's concept. A pair of mini-rovers—Vikings on wheels—could range up to 5 kilometers per day on the martian surface, returning photographs and scientific data via a mother orbiter. Rovers would carry out systematic geological traverses across varied terrain.

signed to mimic to some extent the role of a field geologist making traverses over the countryside: interesting rock samples would be collected (after being examined through the television system by Earth-based geologists), documented, and placed in special containers, in the same way that the Apollo astronauts collected and documented samples on the Moon. A long, organized traverse of this kind could be carried out with the full organization and methodology of scientific exploration used in the Arctic or Antarctic or, even earlier, during the exploration of previously unknown continental areas. Thus there is the potential for, and on some time scale a probability of, a golden age of geographical and geological exploration of the surface of Mars.

Sample Return from Mars

Just as the return of samples from the Moon through the Apollo missions was the crowning scientific achievement of the Apollo program, so the benchmark mission for the scientific exploration of Mars will be the return of selected samples of that planet to the Earth, where the full power of terrestrial laboratory analyses can be brought to bear. This possibility was recognized in the early 1970s, and several concepts were developed in order to accomplish this goal. Indeed, the Soviet Union has already returned small selected samples from the Moon by totally automated means. However, sample return from Mars involves enormously greater distances and therefore greater navigation and endurance requirements for the spacecraft, both going and returning. Most serious, Mars does have an atmosphere, and its surface gravity is twice that of the Moon. Therefore, the problems of entering Mars' atmosphere, landing on its surface, returning to orbit about Mars, and exiting to an Earth-return trajectory are all vastly more complex and difficult than in the lunar case.

But just as low-thrust propulsion offers a way to substitute brain power for unavailable rocket power for the next step in exploring Mercury, so does it offer a solution to the problem of sample return from Mars. Ion drive, in particular, may provide a radical improvement in the prospects for Mars sample return because it greatly reduces the weight of rocket stages that otherwise would have to be sent to Mars in order to return a significant sample (Fig. 8.7). The kind of low-thrust system now under engineering development would be capable of returning perhaps 50 to 100 kilograms of Mars samples to Earth. Furthermore, such a system could deliver the sample container into Earth orbit, where it could be placed in a shuttle-launched receiving laboratory for various kinds of chemical and biochemical tests in orbit. As an alternative, the payload could be returned directly to Earth by means of an Apollo-like re-entry.

Such a low-thrust sample-return mission might be scheduled to follow by two or four years the deployment on Mars of automated rovers to traverse the martian surface and collect documented samples. After traversing the martian landscape and collecting samples, such rovers could be directed to a smooth,

Figure 8.7
An artist's version of a possible future Mars mission shows a lander being deployed from a mother solar-sail spacecraft to pick up Mars soil samples gathered by a team of roving robot vehicles. The rovers would have traversed hundreds—even thousands—of kilometers across Mars over the previous several years, collecting interesting rock samples and depositing them in a central canister for pickup and subsequent return to Earth by the solar sail-powered interplanetary shuttle.

safe landing site where they would await the eventual arrival of the sample-return spacecraft. Altogether, a sample-return mission based on low-thrust propulsion and preceded by autonomous, traversing rovers provides the prospect of an enterprise scientifically comparable to that of the Apollo program itself.

One specific result of the Viking mission has been to reduce greatly the concern that terrestrial microbes accidentally transported to Mars might infect that planet and thereby modify perhaps forever any existing biota. For more than a decade, the planet's surface environment has been known to be quite hostile to terrestrial life forms. The Viking discoveries of the unanticipated environmental hostility to organic compounds generally demonstrates an even greater "self-sterilizing" ability of the martian surface than had been imagined. Additionally, the Viking results probably reduce, in the judgment of many, the speculative possibility that martian soil might contain alien life forms that, if transported to Earth, could survive, reproduce in our environment, become infectious, and cause "plagues." The violent reaction of the martian soil to the mere presence of water suggests that our watery planet's environment would be highly toxic to any martian organism. Nevertheless, the vague spectre of "back contamination" deserves and will surely receive a very full public debate when sample-return missions eventually are seriously considered.

If samples can ultimately be returned from Mars, then the precision age-dating techniques applied to the lunar samples can likewise be used to provide an enormously detailed geologic history of Mars, especially if the samples are carefully collected from known stratigraphic and geographic localities by means of automated vehicles. The full range of analytical techniques—geochemical, biological, and organic chemical—can be applied to the samples by exploiting the enormously diversified and growing capabilities of terrestrial laboratories. Questions about the relationship between such processes as weathering on Mars and the present chemical state of the soil should be directly amenable to analysis under laboratory conditions on Earth. Conversely, in the absence of samples, even as heroic (and costly) an effort as the Viking mission is demonstrably of only limited value if we wish to discover the real biological, chemical, and mineralogical (to say nothing of the isotopic) nature of the martian surface materials.

Just when such momentous steps to explore the next frontier for the human race will be carried out is uncertain. But Viking certainly has indicated what can be accomplished when technical skill and popular imagination are combined on such a goal.

Mars and Human Exploration

Will humans ever go to Mars? The answer must surely be "Yes." Whether it will happen by the end of this century is doubtful. It is helpful, however, to view the exploration of the

Solar System on a very long time scale—one that extends well beyond the lifetimes of people living today. We need to take such a view in order to better understand our role in the long human adventure of exploring the Solar System. From that point of view, the manned exploration of Mars is a very important milestone in future history—one that should help us to define the more immediate goals of scientific exploration in the coming decade.

The first efforts toward landing humans on Mars presumably will be in the nature of an adventuresome journey by astronauts or cosmonauts, as was the case with Apollo. It is a formidable trip, requiring much greater total propulsive capability than that for Apollo. There are no development programs under way by the United States or by the Soviet Union that in any way come close to having the capability for carrying out such a mission. However, the development of an automated *interplanetary shuttle* powered by low-thrust propulsion opens the possibility of a manned mission to Mars, with perhaps ten to twenty shuttle flights to the planet and back (Fig. 8.8).

Whether humans will eventually live on Mars, either in scientific outposts or actual colonies, depends to a large extent upon two questions: (1) Can important activities be carried out more conveniently there or on the Moon or on Earth itself? (2) Can the materials necessary to human habitation be derived locally? The results of the Viking mission partially answer the second question, for they confirm earlier expectations of an abundance of ice and probably chemically bound water in the surface materials. From water, oxygen can be made fairly easily. Thus two of the most important expendables needed to sustain human operations are available locally on Mars, in addition to a rather diminished solar energy flux to provide energy. Indeed, some speculative minds have even contemplated the possibility that, at some distant future time, intentional modification of the martian atmosphere might be undertaken in the desire to convert Mars into a more hospitable planet for human utilization (*terraforming*).

Whether terraforming ever becomes a reality or not, it does seem likely that humans will eventually reach Mars. In spite of

Figure 8.8
When humans go to Mars, this could be the way it will happen, according to an artist's concept. Overhead are several solar-sail-powered interplanetary shuttles, each carrying up to 25 tons. They have landed two manned capsules, various rovers, and processors to mine and study the martian surface. Before astronauts would arrive, a nuclear power station also would be landed.

the hostility of the martian soil to terrestrial kinds of microbial life, the stark and lonely landscape revealed by Viking will one day record human tracks. Viking may ultimately have been a stepping stone to the existence there of very advanced forms of life—*Homo sapiens.*

Beginning the Exploration of the Surface of Venus

Venus remains the most exciting Earthlike planet of which we have not yet acquired an overall picture. The technical devel-

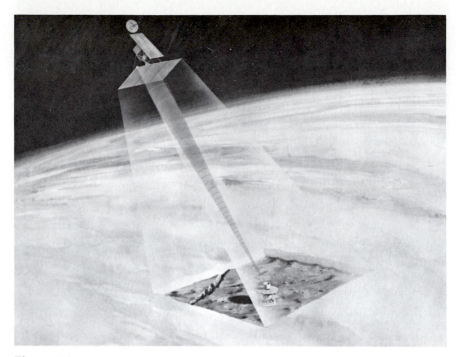

Figure 8.9
Artist's view of a potential mission to Venus, the Venus Orbiting Imaging Radar (VOIR) mission, which would be capable of producing high-resolution images of the surface by imaging radar waves through the dense cloud cover of the planet. Communication with a surface lander is also depicted in this scene.

opment that can make this achievement possible is not propulsion technology but rather imaging radar that recently has been developed to explore the Earth, especially its oceans. Such a radar system could be operated from orbit around Venus to provide detailed maps. It would provide "pictures" of the planet's topography even through its dense clouds. The surface scale would be comparable to that which Mariner 9 provided of Mars, with about 1 kilometer average resolution for most of the planet and a much higher resolution for a small percent (See Fig. 8.9). This would constitute a major scientific bonanza for our understanding of the nature and history of both Earth and Venus.

In addition, further study of the chemical and isotopic composition of the atmosphere and surface must be done to pin down the abundance of volatiles and to compare the histories of surface constituents on Venus with those of Mars and the Earth. Pioneer Venus mission results seem to indicate unsuspected chemical differences between the original volatile components of Mars, Earth, and Venus. Hence we must still deal with the question of where the water went—if it was ever there.

Other Rocky Bodies

The Earthlike planets are not the only rocky bodies in the Solar System. Asteroids are also "rocks," and there are many different orbital classes, probably composed in part of different mineral assemblages. At least some meteorites are believed to have originated in asteroidal collisions that ejected debris into Earth-crossing orbits. Comets also cross Earth's orbit on occasion and may contribute an especially volatile-rich class of meteorites (carbonaceous chondrites) to the diversity of rocks that fall from the sky.

Indeed, many small objects repeatedly cross Earth's orbit and eventually collide with the watery planet. Most are burned up. In 1908, what was probably a fragment of an old comet, about 40–50 meters in diameter, entered the Earth's atmosphere—fortunately over an uninhabited area of Siberia—and exploded with the estimated force of a 10-megaton hydrogen bomb. Trees were knocked down over an area of 40 kilometers in diameter.

As the world's population expands to utilize so much of the available land area, the hazard constituted by other members of that "invisible" band of our closest planetary neighbors should not be ignored, especially since space technology can now provide the means to search out many of our small companions (Fig. 8.10). Indeed, within a few decades it may become possible to change the orbit of a small Earth-crossing asteroid with a potential for future collision. Or it may become possible to capture an Earth-crosser and bring it into the Earth-Moon system for eventual exploitation of its iron or other mineral resources.

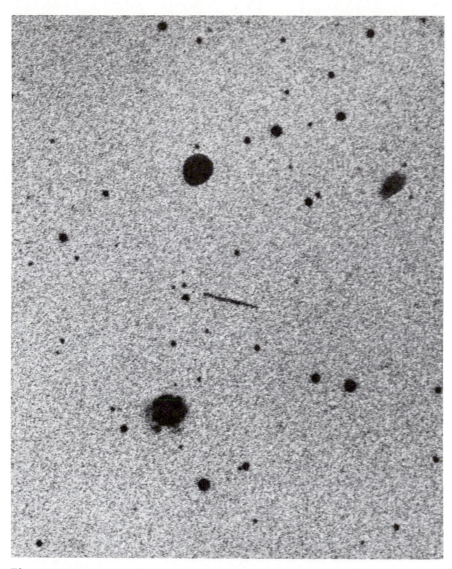

Figure 8.10
Asteroid 1976UA, which came within 750,000 miles of Earth, is shown in this negative print. Such objects are more abundant than realized, and are important to planetary science, geology, and, possibly, human circumstances. This photograph was acquired with the 46-centimeter Schmidt telescope, Hale Observatories. [Courtesy of Eleanor F. Helin.]

How can we learn what asteroids and comets are really like? How can we determine if they are connected to the processes that shaped the inner Solar System or if they are responsible for the origin and evolution of the volatiles of Earth and its neighbors? How can we set the stage for eventual utilization and control of these objects?

The development of a new transportation technology—low-thrust propulsion—offers the heretofore unavailable capability of making a rendezvous with these fast-moving objects. Just as low-thrust propulsion would greatly increase the effectiveness of a Mars sample return or a Mercury orbiter mission, so does it also bring *in situ* measurements of some comets and asteroids within our technological grasp. In fact, low-thrust propulsion would even make it possible to acquire a piece of a comet or an asteroid and return it to Earth for analysis! To catch a comet—what an extraordinary possibility! So thus will asteroids, and especially comets, enter into the realm of scientific reality. And so thus will the family history of the Earthlike planets be completed.

Other rocky bodies are known to orbit Jupiter, Saturn, Uranus, and Neptune. These objects probably formed under quite different conditions from those that gave birth to the Earthlike planets. Yet their compositions of silicates and volatiles must reflect in part processes similar to those that operated during the formation of the inner Solar System. Indeed, surface conditions on the four large moons of Jupiter may still be analogous in some ways to those that prevailed during the first few hundred millions of years on some of the Earthlike planets. In addition, Saturn's large satellite Titan has a moderately dense atmosphere of methane and hydrogen and exhibits a surface environment that is conceivably similar to an early martian or even terrestrial environment.

Voyagers 1 and 2 completed a full reconnaissance of the four large moons of Jupiter. The first glimpses of those planet-sized satellites of Jupiter have been astounding (Figs. 8.11–8.19).

Figure 8.11 *(facing page)*
This view of Io was taken from a range of 862,200 kilometers by Voyager 1. Circular features and irregular depressions are of internal origin and very probably volcanic. Bright irregular patches, possibly solid sulfur dioxide, mask the surface detail in some areas.

Figure 8.12 *(below)*
Voyager 1 acquired this image of Io on March 4, 1979, about eleven hours before closest approach, at a distance of about 490,000 kilometers. An enormous volcanic explosion can be seen silhouetted against dark space over Io's bright limb. The brightness of the plume has been increased by the computer, as it is normally extremely faint. At this time solid material had been thrown to an altitude of about 160 kilometers. This requires an ejection velocity from the volcanic vent of about 1800 kilometers per hour, with material reaching the crest of the fountain in several minutes. The vent area is a complex circular structure consisting of a bright ring about 300 kilometers in diameter and a central region of irregular dark and light patterns.

Figure 8.13
This picture of Europa, the smallest Galilean satellite, was taken from a distance of about 2 million kilometers by Voyager 1 and is centered at about the 300° meridian. The bright areas are probably ice deposits, and the dark areas may be the rocky surface or areas with a more patchy distribution of ice. The most unusual features are the systems of long linear structures that cross the surface in various directions. Some of these linear structures are more than 1000 kilometers long. They may be fractures or faults that have disrupted the otherwise extremely smooth surface.

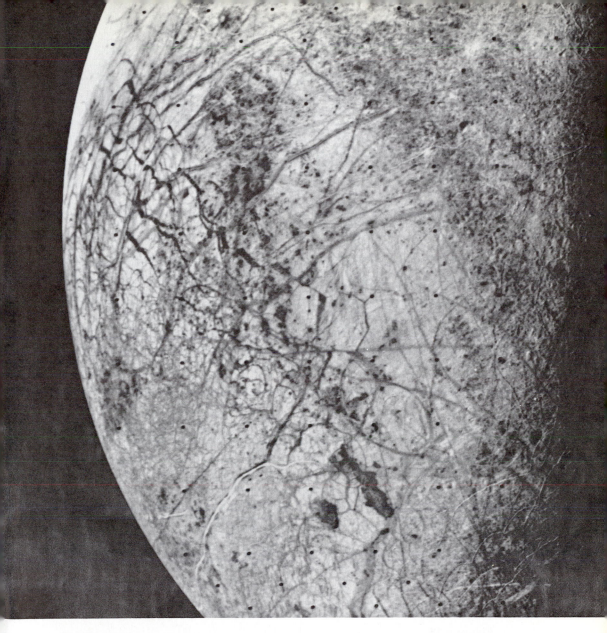

Figure 8.14

This image of Europa was taken by Voyager 2 on July 9, 1979. The area shown is about 600 by 800 kilometers. This image was taken along the evening terminator, which best shows the surface topography of complex narrow ridges, seen as curved bright streaks, 5 to 10 kilometers wide and hundreds to thousands of kilometers in length. A few features are suggestive of impact craters but are rare, indicating that the probably icy surface is still (or was in the geologically recent past) active, perhaps warmed by tidal heating like Io. The larger icy satellites, Callisto and Ganymede, are evidently colder, with much more rigid crusts still bearing the scars of ancient impact craters. The complex intersecting of dark markings and bright ridges suggests that the surface has been fractured and darker material from beneath has welled up to fill the cracks.

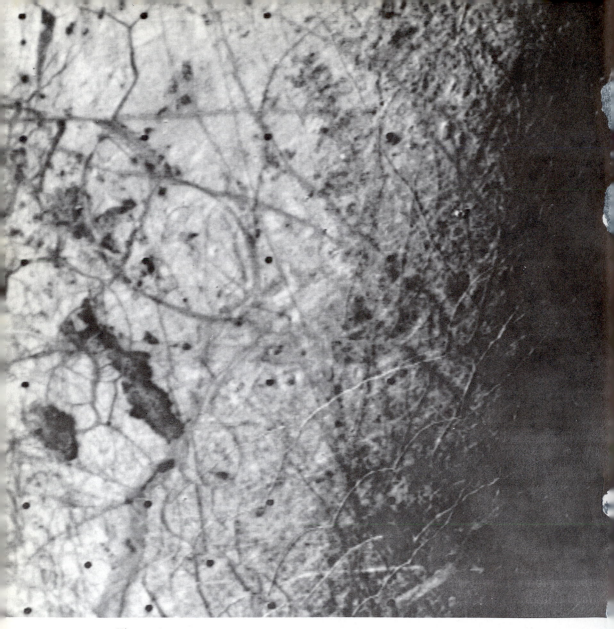

Figure 8.15
This image of Europa, smallest of Jupiter's four Galilean satellites, was acquired by Voyager 2 on July 9, 1979, from a range of 241,000 kilometers and shows a portion of the area seen in Figure 8.14. Europa, the brightest of the Galilean satellites, has a density slightly less than that of Io, suggesting it has a substantial quantity of water. Scientists previously speculated that the water must have cooled from the interior and formed a mantle of ice perhaps 100 kilometers thick. The complex patterns on its surface suggest that the icy surface was fractured and that the cracks filled with dark material from below. Very few impact craters are visible on the surface, suggesting that active processes on the surface are still modifying Europa. The tectonic pattern seen on its surface differs drastically from the fault systems seen on Ganymede, where pieces of the crust have moved relative to each other. Europa's crust evidently fractures, but the pieces remain in roughly their original position.

Figure 8.16
This picture of Ganymede was taken by Voyager 1 from a distance of 2.6 million kilometers. Ganymede is Jupiter's largest satellite with a radius of about 2600 kilometers, about 1.5 times that of Earth's Moon. Ganymede has a bulk density of only approximately 2.0 grams per cubic centimeter, much less than that of the Moon. Therefore, Ganymede is probably composed of a mixture of rock and ice. The large dark regions in the northeast quadrant, and the white spots, resemble features found on the Moon—specifically, mare plains and impact craters. The long white filaments resemble rays associated with impacts on the lunar surface. The various colors of different regions probably represent differing surface materials.

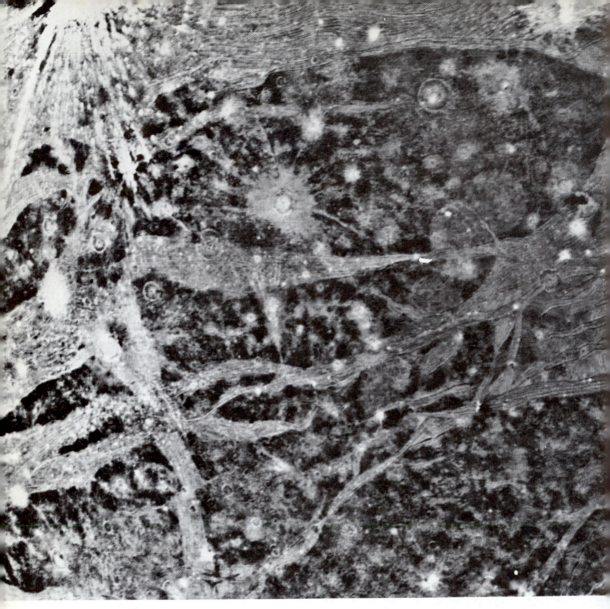

Figure 8.17

Ganymede, seen by Voyager 1 from a range of 246,000 kilometers. The center of the pictures is at 19° south latitude and 356° longitude; the picture is about 1000 kilometers across. The smallest features seen are about 2.5 kilometers across. The surface displays impact craters, many of which have extensive bright ray systems. The craters lacking ray systems are probably older than those showing rays. Bright bands traverse the surface in various directions and contain an intricate system of alternating linear bright and dark lines, which may represent deformation of the crustal ice layers. These lineations are particularly evident near the top of the picture. A bright band trending in a north-south direction in the lower left-hand portion of the picture is offset along a bright line. This offset is very probably due to strike-slip faulting. Two light circular areas in the right upper center of the picture may be the scars of ancient impact which have had their topographic form erased by flowage of the crustal icy material.

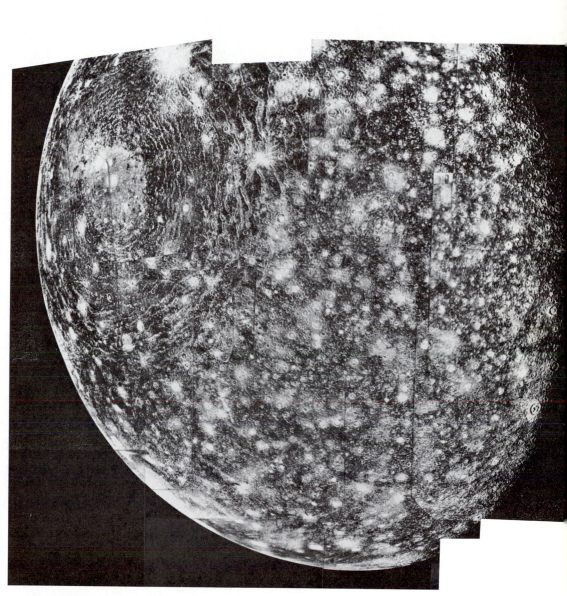

Figure 8.18

Photomosaic of Callisto, imaged by Voyager 1 from a range of 202,000 kilometers. Callisto is the darkest and outermost of the Galilean satellites of Jupiter. The surface has been heavily cratered by ancient meteorite impacts and is probably the oldest surface of the Galilean satellites. Many of the craters display bright ray systems similar to those on Earth's Moon. The large bright spot near the upper left corner of the picture, near Callisto's limb, is an impact basin about 300 kilometers in diameter.

Figure 8.19
High-resolution picture of the multiringed basin on Callisto seen in Figure 8.18. The complicated circular structure is similar in some ways to the large circular impact basins that dominate the surface of the Moon, Mars, and Mercury. This multiringed basin on Callisto consists of a light-floored central basin some 300 kilometers in diameter surrounded by at least eight to ten discontinuous, rhythmically spaced ridges. No radially lineated ejecta can be seen.

Active volcanism was actually observed reworking the bright orange surface of Io, the innermost of the four large moons of Jupiter. And the primary substance composing the associated lavas may be sulfur, not silicate basalt. In contrast, Callisto—the outermost large moon—exhibits a dark, heavily cratered surface that probably still preserves some of the scars from the terminal period of heavy bombardment. Yet Europa, next out from volcanic Io, shows no evidence of volcanism or sulfur and practically no craters either! Europa is apparently the smoothest planetary body yet encountered in our Solar System. Finally, Ganymede, orbiting between Europa and Callisto, displays a marvelously variegated surface in which large pieces of dark, cratered, Callisto-like crust are intertwined with younger icy crustal material apparently recording an ancient period shortly after the end of heavy bombardment when internal convection (mostly of water in this case) partly destroyed the original Callisto-like surface. Ganymede exhibits abundant faults, horizontally offsetting adjacent blocks (i.e., *strike-slip* faults). Only on Earth had such phenomena been observed prior to the Voyager observations. It may well be that Ganymede, which is probably half water in composition, really records a process analogous to plate tectonics on Earth.

Ultimately, the bombardment histories of these moons of Jupiter will be related to that of the Earthlike planets, perhaps providing crucial evidence whether the early bombardment and its abrupt cessation occurred throughout the Solar System.

Earth and the Future

What about the Earth itself? Are its volatiles alien or are they part of the same stuff from which the Earth itself was initially formed? And how about us? Are we, too, of alien origin, since we are made of those same volatiles? How did life start here if, indeed, it really originated on the Earth? What is the connection, if any, between the onset of the glacial period and the development of intelligent animals from primates? The whole of human history is intimately tied into climatic change.

The polarity of our magnetic field has reversed itself many times throughout history, and at times there has been no magnetic field at all. When there is no magnetic field, there is little shielding of the Earth's surface from high-energy radiation from the Sun and the cosmos. What has been the effect of brief anomalous magnetic periods upon evolution and the development of the animals and plants of our world? Has it temporarily accelerated the number of mutations? And what really controls the concentration of minerals and ores in the Earth's crust? What else besides the colliding of tectonic plates is involved in causing the very nonuniform and extraordinary distribution of useful elements and chemicals? Answers to these and other important questions about the Earth may be obtained by studying our neighboring planets.

The Search for Extraterrestrial Intelligence

Are we the only intelligent beings in the Universe? Is the formation of a planet with a watery envelope—which gave rise to life as we know it—an extremely rare event in the Universe? There are *billions* of stars like the Sun within our own Galaxy. And nothing that we have learned so far about the history of our Sun or of the Earth or its inhabitants suggests that their origin and evolution are unique. Rather, Earth, life, and intelligent inhabitants all seem to be a reasonable outcome of stellar processes that must operate throughout the Universe. It is difficult to avoid the notion that the development of inhabited planetary systems must be a fairly common occurrence throughout the enormous expanse of our galaxy. The detection of evidence for the existence of extraterrestrial intelligence would be the most significant scientific discovery imaginable.

If there are technological societies—perhaps much more advanced than our own—orbiting other stars elsewhere in the galaxy, can we communicate with them? We shall not know without trying to listen for communications from the cosmos. Indeed, a most challenging observational task ahead is to search for radio signals of artificial origin from outside our own Solar System.

Over the past two decades, the exploration of our inner Solar System has been an exploration back in time to the origins of our planet, to the origins of our water, to the origins of our organic material, and therefore to the origin of human life and intelligence. The next phase involves not only continuing the exploration of our planetary neighbors and ourselves, but also a search outward well beyond the Solar System for our analogs who may exist elsewhere and who would constitute the community of our galaxy.

APPENDIX

GUIDE TO PLANETARY PHOTOGRAPHY

The purpose of this appendix is to supply source information for those who may wish to obtain copies of photographs that appear in this book. Presented here in tabular form are (1) the figure numbers used in this book, (2) a brief description of each subject, (3) the identification number to use when ordering a photograph, and (4) the source. Look up the source number under the heading Source Information, which follows the table.

Most of the photographs are available at nominal cost from the National Space Science Data Center (address given under Source Information). Some photographs used in this book were specially processed for us and are not available in exactly the same form, in which case we have given identification numbers for photographs most like those used in the book. A few photographs, not generally available, were provided by individuals, including the authors. Some figures are actually enlargements of a part of an original photograph; the identification numbers for these are followed by (p).

Figure	Description	Identification number	Source
1.1	Apollo 17 Earth: Global View	AS17-148-22727H	1
1.3	Lick Observatory: Quarter Moon Mosaic	LOP L-9	2
1.4	Mariner 9 Mars: 3-frame mosaic	M9 DAS 12994242	3
		M9 DAS 12994312	3
		M9 DAS 12994382	3
1.5	Mariner 10 Venus: UV Clouds	JPL P-14400	6
1.6	Mariner 10 Mercury: Incoming Hemisphere	AOM Figure 18	3
	Outgoing Hemisphere	AOM Figure 19	3
2.1	Lick Observatory: Full Moon	LOP L-4	2
2.2	Apollo 16 Moon: Global View	AS16-3029M	1
2.6	Mariner 10 Mercury: South Polar Mosaic	AOM Figure 20	3

Figure	Description	Identification number	Source
2.7	Mariner 6 Mars: Equatorial Mosaic	M6:6N09-6N24	3
2.9	Mariner 9 Mars: Valles Marineris Mosaics	(M. Malin)	
2.10	Viking Mars: South Polar Mosaic	JPL 211-5393(p)	6
2.13	Venera Venus: Lander Photographs	(C. P. Florensky, Academy of Sciences, U.S.S.R.)	
3.1	USN Earth: Antarctica Labyrinth	USN 14NOV59 MCMANT 123 003F33 006	7
3.2	Viking Mars: Fretted Terrain	JPL P-18086	6
3.3	Apollo 17 Moon: Craters	AS17-150-22947H	1
3.4	Apollo 17 Moon: Euler Crater	AS17-2923M	1
3.5	Apollo 17 Moon: Regolith	AS17-145-22165H	1
	Viking Mars: Regolith	L1.C1.AM.Q3+Q4(p)	5
	Venera Venus: Regolith (Venera 9)	(C. P. Florensky, Academy of Sciences, U.S.S.R.)	
	Malin Earth: Regolith (Iceland)	(M. Malin)	
3.8	Apollo 16 Moon: King Crater	AS16-122-19580H	1
3.9	Viking Mars: Arondus Crater/Cydonia	JPL P-17871	6
3.10	Lunar Orbiter Moon: Orientale Basin	LO IV 187M	4
3.11	Gault: Mare Exemplum	(D. Gault)	
3.16	Apollo 17 Moon: Dawes Crater	AS17-2762P	1
3.17	Apollo 17 Moon: Boulder Tracks	AS17-144-21991H	1
3.18	Viking Mars: Avalanche Chutes	VO1:90A63	5
3.19	Viking Mars: Landslide	JPL P-16952	6
3.20	Mariner 10 Mercury: Crater Mass Movement	M10 FDS 529024	3
3.24	Mars Wind Streaks	VO1:56A20,20A54,10A39	5
		VO2:45B46,45B60	5
3.25	Greeley: Wind Tunnel Streaks	(R. Greeley)	
3.26	Viking Mars: Dunes in Crater	VO2:510B88	5
3.27	Mariner 9 Mars: Layered Deposit	M9 DAS 9231184	3
3.28	Viking Mars: Chryse Planitia (oblique)	VO1:4A06-4A10	5
3.29	Mariner 9 Mars: Mangala Vallis	M9 DAS 6822728	3
3.30	Mariner 9 Mars: Mangala Vallis Hi Res.	M9 DAS 9628649	3
3.31	Mars: Runoff Channel	M9 DAS 6606708	3
3.32	Mars: Fretted Channel	VO1:529A22	5
4.1	Apollo 16 Moon: Alphonsus	AS16-2478M	1
4.2	Apollo 15 Moon: Hadley Rille	AS15-0587M	1
4.3	Apollo 15 Moon: Euler (Imbrium) Flows	AS15-1557M	1
4.5	Model: "Lithosphere"		8
4.6	Landsat Earth: Landsat mosaic—Iceland	1392-12185	7
		1392-12191	7
		1426-12070	7
		2494-11503	7
4.8	Landsat Earth: Altaplano Volcanics	2148-13415	7
		2148-13421	7

Figure	Description	Identification number	Source
4.9	Landsat Earth: Southern California	1090-18012	7
4.10	Landsat Earth: Altyn Tagh Fault	1074-04253	7
4.12	Glass Mountain	(R. Greeley)	
4.13	Skylab Earth: Mt. Etna	S3-87-355	7
4.15	Goldstone Venus: Venus Trough	JPL 331-4730A	6
		JPL 331-4735A	6
		JPL 331-4738B	6
		JPL 331-4739A	6
4.16	Viking Mars: Lava Flows	VO1:56A14	5
4.19	Earth: Leveed Channel	(R. Greeley)	
4.20	Moon: Leveed Channel	(R. Greeley)	
4.21	Elongate Vents	(R. Greeley)	
4.22	"Star" Vents	(R. Greeley)	
4.23	Moon and Mares: Mare Ridges	AS17-0458M	1
		M9 DAS12901348	3
4.24	Moon: Lunar Domes	LO V 185M (l)	4
		LO V 182 (r)	4
4.25	Mariner 10 Mercury: Caloris Zoom	M10 FDS 198	3
		M10 FDS 529057	3
		M10 FDS 528998	3
4.26	Apollo Moon: Humboldt	AS15-2512M	1
4.27	Earth: Mauna Loa/Mauna Kea	USN 12NOV64 HAI 0066 22/415	7
4.29	Earth: Kilauea Caldera	USN 1NOV54 HAI 0010 381F	7
4.30	Viking Mars: Olympus Mons	VO1:646A28	5
4.31	Mariner 9 Mars: Lava Tube	M9 DAS 5492408	3
4.32	Viking Mars: Ascreas Mons Caldera	VO1:90A50	5
4.33	Viking Mars: Tharsis Tholus	VO1:853A23	5
4.34	Mariner 9 Mars: Hecates Tholus	M9 DAS 13496293	3
		M9 DAS 13496363	3
4.35	Mariner 9 Mars: Elysium Mons	M9 DAS 1346083	3
	Apollo Earth: Emi Koussi	AS7-5-1621 H	7
4.36	Lunar Orbiter: Small Shields	LO IV 187H2	4
4.37	Viking Mars: Graben	VO1:63A07	5
4.38	Viking Mars: Chaotic Terrain	JPL 211-5821 (p)	6
4.39	Mariner 10 Mercury: Discovery Scarp	AOM Picture 11-27	3
4.40	Topographic Map of Venus	(Harold Masursky, USGS, 601 East Cedar Ave., Flagstaff AZ 86001)	
5.5	Apollo Moon: Breccia within Breccia	NASA S-76-29484	1
5.7	Apollo Moon: Terrae (oblique)	AS16-0847M	1
5.8	Apollo Moon: Fra Mauro Formation	AS12-52-7597 H	1
5.9	Lunar Orbiter: Orientale Flow Ejecta	LO IV 172H2	4
5.10	Apollo Moon: Hadley Rille	AS15-0587M	1
5.11	Apollo Moon: Cayley Formation	AS16-0440M	1

Figure	Description	Identification number	Source
6.2 (left)	Mariner 10 Mercury: Santa Maria Rupes	M10 FDS 27448	3
6.2 (right)	Mariner 10 Mercury: Northern ICP	M10 FDS 27252	3
6.3 (top)	Mariner 10 Mercury: Old Basin Stereo	M10 FDS 27387 (r)	3
		166667 (l)	3
6.3 (middle)	Mariner 10 Mercury: Basin Superposition	M10 FDS 27301 (r)	3
		166649 (l)	3
6.3 (bottom)	Mariner 10 Mercury: Ridge	M10 FDS 27403 (r)	3
		166661 (l)	3
6.4	Mariner 10 Mercury: Radial Ejecta	M10 FDS 166	3
6.5	Mariner 10 Mercury: Caloris Basin Mosaic	AOM Figure 21	3
6.6	Mariner 10 Mercury: Caloris Sculpture	M10 FDS 193	3
6.7	Mariner 10 Mercury: Smooth-Hummocky	M10 FDS 072	3
6.8	Mariner 10 Mercury: Caloris Fractures	M10 FDS 126	3
6.9	Mariner 10 Mercury: Hilly & Lineated Terrain	M10 FDS 27374	3
6.10A	Mariner 10 Mercury: North Polar Mosaic	AOM Photomosaic 1-B	3
6.10B	Mariner 10 Mercury: South Polar Mosaic	AOM Photomosaic 15-B	3
		AOM Photomosaic 15-C	3
7.3A	Mariner 6 Mars	6N21	3
7.3B	Viking Mars	V01:618A23-618A28 (p)	5
7.5	Viking Mars: Argyre and Happy Face	JPL P-17022	6
7.6 (top)	Mariner 9 Mars: Intercrater Plains	M9 DAS 11620145	3
7.6 (bottom)		M9 DAS 8909609	3
7.7 (top)	Mariner 9 Mars: Ganges Chasma Low Res	M9 DAS 7614498	3
7.7 (bottom)	Mariner 9 Mars: Ganges Chasma Hi Res	M9 DAS 7017619	3
7.8	Viking Mars: Capri Chasma	JPL P-16983	6
7.9	Viking Mars: Valley Networks	JPL 211-5207	6
7.11	Viking Mars: Chryse Channels	VO1:520A27-520A34	5
7.12	Viking Mars: Chryse Landing Site	VO1:27A35	5
7.13	Viking Mars: Ceraunius and Uranius Tholus	JPL 211-5639	6
7.14	Viking Mars: Apollonaris Patera	JPL 211-5213	6
7.15	Viking Mars: North Polar Dunes	VO2:58B21-58B34 (p)	5
7.16	Mariner 9 Mars: Variable Surface Features	M9 DAS 5707273 (l)	3
		M9 DAS 8585259 (r)	3
7.17	Viking Mars: Yardangs	VO2: 44B37	5
7.18	Viking Mars: Dust Storm	JPL P-18795	6
7.19	Viking Mars: First Lander Picture	JPL P-17043	6
7.20	Viking Mars: Second Lander Picture	JPL P-17681	6
7.21	Viking Mars: Lander 2 General	JPL P-17683	6
8.1 (upper left)	Mariner 10 Mercury: Cratered Terrain	M10 FDS 27256	3
8.1 (lower left)	Mariner 10 Moon: Cratered Terrain	M10 FDS 2669	3
8.1 (upper right)	Viking Orbiter Mars: Cratered Terrain	VO1:793A10	3
8.2A	Apollo 15 Moon: Mare Ridges	AS15-2071M	1
8.2B	Viking Mars: Mare-like Ridges	V01: 87A17	5
8.2C	Mariner 10 Mercury: Mare-like Ridges	M10 FDS 095	3
8.3A	Earth: Graben	(G. McGill)	
8.3B	Viking Mars: Graben	V02: 7B29	5

Figure	Description	Identification number	Source
8.3C	Apollo 15 Moon: Graben	AS15-1679M	1
8.3D	Mariner 10 Mercury: Graben	M10 FDS 106	3
8.4	Ion Engine	JPL P-18552AC	6
8.5	Ion Engine	JPL 383-5087	6
8.6	Mars Rover	JPL P-16919C	6
8.7	Mars Sample Return	JPL P-17055Bc	6
8.8	Mars Manned Base	JPL P-17055Ac	6
8.9	VOIR Mission: Artist's Concept	JPL P-17131Bc	6
8.10	Asteroid 1976 UA	(E. F. Helin)	
8.11	Voyager I Io: Overview	JPL P-21209	6
8.12	Voyager I Io: Eruption	JPL P-21305C	6
8.13	Voyager I Europa: Overview	JPL P-21208C	6
8.14	Voyager II Europa: Close-up	JPL P-21765BW	6
8.15	Voyager II Europa: Close-up	JPL P-21766BW	6
8.16	Voyager I Ganymede: Overview	JPL P-21207C	6
8.17	Voyager I Ganymede: Close-up	JPL P-21266	6
8.18	Voyager I Callisto: Overview	JPL P-21282	6
8.19	Voyager I Callisto: Close-up	JPL P-21233	6

Color plates			
1	Viking Mars: Lander View of Utopia	JPL P-18296C	6
2	Viking Mars: Frost Photo	JPL P-21873C	6
3	Voyager: Jovian System Collage	JPL P-21631C	6
4	Voyager Io: Pizza Pie	JPL P-21457C	6
5	Voyager Io: Color Eruption Plume	JPL P-21305C	6
6	Voyager Europa: Global Mosaic	JPL 260-686	6
7	Voyager Ganymede: Global Mosaic	JPL 260-671	6
8	Voyager Callisto: Global Mosaic	JPL 260-450	6

SOURCE INFORMATION

1. Apollo photographs of the Moon are available from National Space Science Data Center (NSSDC):

> National Space Science Data Center
> Code 601, Goddard Space Flight Center
> National Aeronautics and Space Administration
> Greenbelt, Maryland 20771

   AS means Apollo Saturn and is followed by the mission number. Subsequent numbers uniquely identify the photograph.
   H means Hasselblad 70-mm was the original format of the picture.

M   means Metric Camera 5-inch was the original format of the picture.

P   means Panoramic Mapping Camera (5″ × 48″) was the original format.

NASA S   means a NASA Special photograph that may be available from NSSDC.

2. Lick Observatory photographs may be purchased from:

Lick Observatory OP
University of California
Santa Cruz, California 95064

L   means lunar photograph, and the number that follows is the catalogue number.

3. Mariner photographs of Mercury, Venus, and Mars are available from NSSDC (see item 1 for address).

M6   means Mariner 6 Mars flyby
The code number 6N_ _ means Mariner 6, Near Encounter, and the numbers that follow refer to the individual pictures uniquely. Order *Maximum Discrimination* version.

M9   means Mariner 9 Mars Orbiter
DAS means Data Acquisition System. The number following DAS uniquely identifies the photograph.
Order *Vertical* or *Horizontal High Pass Filtered* version.

M10   means Mariner 10 Venus/Mercury flyby
FDS means Flight Data System. The number following FDS uniquely identifies the photograph.
Order the *IPL* version.

AOM   mean *Atlas of Mercury* prepared from Mariner 10 images. Specify figure, photomosaic, or picture, and the number to identify the photograph uniquely.

4. Lunar Orbiter photographs of the Moon are available from NSSDC (see item 1 for address).

LO   means Lunar Orbiter, the Roman numeral refers to the mission (I−V) and the frame number. M or H1−H3) refers to medium or high-resolution frame and must be specified.

5. Viking photographs of Mars are available from NSSDC (see item 1 for address).

VO_   means Viking Orbiter; number refers to spacecraft 1 or 2.

_A_   refers to orbit number, spacecraft (A = 1, B = 2) and frame number. Specify all numbers, and order MTVS NGF version.

L1.C1.AM.Q3+Q4 is a Viking Lander 1 view, a computer-generated photomosaic.

6. JPL photographs are available from NSSDC (see item 1 for address) except as noted below (JPL 331).

JPL   means a Jet Propulsion Laboratory Press Release photograph.

JPL 211   means a Viking Orbiter mosaic.

JPL 331   means JPL internal negative number; not available from NSSDC.

7. Photographs of the Earth may be purchased from:

User Services Unit
EROS Data Center
Sioux Falls, SD 57198
(605) 594-6511, ext. 151

USN   U. S. Navy photographs that may be available from EROS. USN is followed by the date the photographs were taken and the general location (MCMANT = McMurdo Base, Antarctica; HAI = Island of Hawaii). The number identifies the mission and frame number (italics). These frames may not be generally available.

Landsat photographs are designated by 9-digit identification numbers the first digit being the spacecraft number. Landsat Multispectral scanner images used in this book are Band 4 (green) black and white or Bands 4, 5, and 7 (green, red, and IR2) color composites. These composites may not be available from EROS in the form seen in the book.

Skylab photographs of the Earth are designated by S (Skylab) followed by the mission number (1, 2, or 3) and the photograph number.

Apollo Earth photographs are identified by AS mission number (7, 9) and the frame number. H refers to 70-mm Hasselblad format.

8. "Lithosphere" photographs can be purchased from:

Dick Thomas
Lithosphere
3205 Alcazar, NE
Albuquerque, NM 87110

9. The major published collections of orbital photography of the Earthlike planets are listed as references at the end of the appropriate chapters and may be consulted for additional photography.

INDEX

Pioneer Venus missions, 102, 190, 353
Pit craters, 172
Plagioclase feldspars, 130, 207, 220
Plains
    of lunar highland, 215–218
    on Mars, 43, 181, 186–187, 292–306, 328–329
    on Mercury, 34–35, 128, 130, 162–163, 187, 237–242, 251–253, 257–264
    nonvolcanic, 162–163, 215–218
    volcanic, 152–162, 219–227, 292–306
    See also Maria regions, lunar
Planets
    atmospheres of, 305
    comparison of, 246, 256–257, 262–264, 321–367
    formation of, 4–9, 124
    See also individual planets
Plate boundaries, 133–141
Plate collision, 136–137
Plate tectonics, 57, 133–151, 189, 304, 332–333
Polar regions
    of Earth, 48–50, 65–68
    of Mars, 45–47, 110–112, 278–279, 306–311
Potassium, 7, 124, 198, 207
Potassium/argon dating, 204–207
Primary atmosphere, 5–7, 11
Prograde rotation, 18
Pyroxene, 130, 220

Radio emissions, from outer space, 340, 366
Radioactive elements, 7, 124, 198
Radioactive isotopic dating, 204, 206, 220
Rainier, Mt., 146
Rare earth elements, 207
Rarefaction wave, 72
Refractory elements, lunar, 131
Regolith, 74–75, 80
Retrograde rotation, 18, 54–56
Rhyolite, 145

Ridges
    on Mars, 284, 287, 304, 307, 310
    on Mercury, 163, 183, 186–187, 249–254, 257–258
    on Moon, 158–162, 210
    on terrain map, 201
Rift, 136
Rilles, lunar, 128, 129, 158–159, 181, 216
Rook Mountains (Moon), 82, 84, 214
Rotation
    of Moon, 27
    of Venus, 18, 54–56
Rubidium/strontium dating, 204–207
Runoff channels, martian, 114, 117

Salt weathering, 68
Saltation creep, 44, 102, 104
San Andreas Fault, 141, 142
Santa Maria Plains (Mercury), 240
"Sapping" erosion, martian, 287, 290
Saturn, 355
Scarps
    diagram of, 92
    on Mars, 98, 284, 287, 310
    on Mercury, 36, 187, 240, 244, 246–247, 250–253
    on Moon, 214
Schiaparelli, Giovanni, 272
Secondary atmosphere, 7
Secondary impact craters
    and dating techniques, 88
    description of, 73
    on Mars, 276
    on Mercury, 34, 78, 239, 244
    on Moon, 218
Seismic disturbances. *See* Faults
Shield volcanoes
    description of, 158
    diagram of, 155
    on Earth, 166–170
    on Mars, 170–178, 297–306
    on Moon, 178–179
Siderophilic elements, 198
Silicate minerals
    in Earth's crust, 7
    in Earth's mantle, 9, 136
    on Mercury's surface, 236–237